ISBN 978-1-330-26044-9
PIBN 10004706

English
Français
Deutsche
Italiano
Español
Português

www.forgottenbooks.com

Mythology Photography **Fiction**
Fishing Christianity **Art** Cooking
Essays Buddhism Freemasonry
Medicine **Biology** Music **Ancient**
Egypt Evolution Carpentry Physics
Dance Geology **Mathematics** Fitness
Shakespeare **Folklore** Yoga Marketing
Confidence Immortality Biographies
Poetry **Psychology** Witchcraft
Electronics Chemistry History **Law**
Accounting **Philosophy** Anthropology
Alchemy Drama Quantum Mechanics
Atheism Sexual Health **Ancient History**
Entrepreneurship Languages Sport
Paleontology Needlework Islam
Metaphysics Investment Archaeology
Parenting Statistics Criminology
Motivational

[AS]TRONOMY

[WIT]HOUT A TELESCOPE

...ide to the Constellations, and Introduction to the
...tudy of the Heavens with the Unassisted Sight.

BY

. WALTER MAUNDER, F.R.A.S.

The Royal Observatory, Greenwich...

ILLUSTRATED BY STAR MAPS AND KEY DIAGRAMS.

London
W. THACKER & CO., ...
THACKER, SPINK & CO., CALCUTTA AND SIMLA.

·ASTRONOMY·
WITHOUT A TELESCOPE

A Guide to the Constellations, and Introduction to the Study of the Heavens with the Unassisted Sight.

BY

E. WALTER MAUNDER, F.R.A.S.

Author of "The Royal Observatory, Greenwich; Its History and Work."

ILLUSTRATED BY STAR MAPS AND KEY DIAGRAMS.

London:

W. THACKER & CO., 2, CREED LANE, E.C.
THACKER, SPINK & CO., CALCUTTA AND SIMLA.

1904.

PREFACE.

THE last quarter of a century has seen a striking increase in the size and efficiency of astronomical telescopes. These have opened up new fields of great interest; but in the widespread attention which they have attracted there is some danger lest it should be forgotten that there are fields for which the primary instrument of all, the unassisted human eye, is still available, and to which it alone is adapted, fields which even yet are far from being exhausted.

I owe my own first recognition of the extent of the work in astronomy which there is still to be done without a telescope to the sight of the Zodiacal Light whilst passing through the Red Sea. The more I studied that object, the more I was impressed with its mysteriousness, and it was in the first place, in the hope of enlisting observers in its study, that I commenced in 1900 a series of short papers in KNOWLEDGE upon the Zodiacal Light and other objects which like it need no optical assistance for their examination. The subject has grown in the writing to far larger dimensions than I had in the least anticipated at first, and no doubt it might have been expanded to a much greater length still without exhausting it. I hope, however, that what I have written may prove suggestive, and perhaps helpful to those who, though they possess no great equipment, yet have a love for the first and grandest of all the sciences, and a desire to take a practical share in it. Whilst, even in these days, there are still many who delight to see spread out before them night after night the glories of the heavens, and—

> " To read the page,
> Where every letter is a glittering world,"

and to whom that high contemplation never fails to bring a " certain joyful calm." It may be that these will not be unwilling to read once again of the shining ones with which they are so familiar, and perhaps in some cases will be glad to be led from the pleasure of admiration alone to the greater pleasure of admiration linked with ordered observation.

For the illustrations of this book I am greatly indebted to Miss C. O˙ Stevens, who has prepared the maps for me, and who lent her drawing of the eclipse of 1900; to Miss Bacon, M. Antoniadi, M. Easton, Col. E. E. Markwick, and Mr. H. Keatley Moore, for their permission to use photographs or drawings, and to the Council of the British Astronomical Association for the loan of several blocks.

E. W. M.

St. John's, London, S.E.,
    October, 1902.

CONTENTS.

INTRODUCTION.

SECTION I.

CONSTELLATION STUDIES.

CHAPTER I.

ORIGIN OF THE CONSTELLATIONS.

CHAPTER VIII.

THE SOUTH CIRCUMPOLAR STARS.

SECTION II.

ASTRONOMICAL EXERCISES WITHOUT A TELESCOPE.

CHAPTER I.

THE SUN AND THE SEASONS.

CHAPTER II.

MORNING AND EVENING STARS.

CHAPTER III.

THE MARCH OF THE PLANETS.

SECTION III.

ASTRONOMICAL OBSERVATIONS WITHOUT A TELESCOPE.

CHAPTER I.

METEORS.

CHAPTER II.

THE ZODIACAL LIGHT.

CHAPTER VIII.

STARS BY DAYLIGHT AND THE SUM OF STARLIGHT.

CHAPTER IX.

VARIOUS SKY EFFECTS.

CHAPTER X.

VARIABLE STARS.

CHAPTER XI.

THE COLOURS OF STARS.

ILLUSTRATIONS.

STAR MAPS.

———

INDEX TO THE CONSTELLATIONS.

INTRODUCTION.

OME years ago, when the Sioux Indians were beginning to get restless and to threaten trouble, it was thought expedient by the authorities at Washington to invite some of the discontented chiefs to an interview with their "Great White Father," the President, and, incidentally, to give them a demonstration of the vast resources which they would have to encounter if ever they took up arms against the Federal Government. So they came, and were shown some of the mighty machines which modern engineering has produced, and, in particular, some hundred-ton guns. The monster weapons were duly manœuvred for the red men's benefit. They were loaded and fired, and the Indians were conducted to the ruin which had been the target that they might mark the terrible destruction which the missile

B

had wrought. The Indians looked, but instead of being overwhelmed with astonishment and fear, as their guides had expected, betrayed only a slightly bored indifference. The United States official in charge of the demonstration repeated and emphasized the explanations, when one of the chiefs, with just the faintest ghost of a satirical smile, which was the utmost manifestation of feeling his stoical sense of dignity allowed him, said, pointing to the unwieldy weapon, "You won't come after Indian with that."

It was true! The officials felt its force at once, and the Indians were treated to no more exhibitions of heavy artillery practice. It had been forgotten that the most powerful weapon is not necessarily the most effective for every purpose, and that for some classes of work the great size of an instrument may be a fatal disqualification.

A very similar mistake is sometimes made in regard to astronomy, and has no doubt interfered with the popularity of the science as a pursuit. It is too often assumed that nothing of real interest or utility can be achieved without the possession of telescopes of enormous power and of corresponding cost. The great observatories maintained in various European countries by the State, or founded in America by millionaires, like Lick or Yerkes, have been thought to command a monopoly of the astronomical advances of the future, since they alone possess the telescopes of greatest light-gathering power and most perfect definition.

This view is far from correct. In the first place such an assumption entirely overlooks the fact that

"By the time a refractor of this kind has been erected and equipped, the outlay upon it will have become so large that it would be utter folly to use the instrument for work other than that for which its great power renders it specially fitted. The result of this is that our modern giant telescopes are, with few exceptions, employed, not in doing work which was formerly done by smaller instruments, but in doing work which formerly could not be done at all. Such, for instance, is the bulk of stellar spectroscopic work, including determinations of velocity in the line of sight, the measurement of close double stars, the spectroscopic examination of nebulæ, the discovery of new planetary satellites, and similar matters. We see, therefore, that the establishment of these powerful telescopes has been accompanied by the development of new fields of research, and that the work which was formerly done—and can still be well done—by instruments of moderate size has not been reduced." *

Nor is this all. Not only are the new giant telescopes necessarily devoted almost entirely to work which smaller instruments cannot touch, thus leaving to the latter the observations within their compass, but there are departments of work for which a great refractor is as wholly unsuited as a hundred-ton gun would be for fighting a Red Indian or shooting snipe. Great light-gathering power is not always the most important

* Mr. W. H. MAW, F.R.A.S., Presidential Address to the British Astronomical Association, October 25th, 1899. *Journal of the British Astronomical Association*, Vol. X., No. 1, p. 8.

quality; for some researches broad grasp of field is far more essential, and here the giant telescopes are practically useless.

Prof. E. E. Barnard, in one of his lectures on Astronomical Photography, illustrated this point by showing a photograph of the great nebula in Andromeda, with all the marvellous detail of ring within ring which the photographs of Dr. Roberts and his followers in this field have made familiar to us. Then over this he would place a mask, cutting down the field of view to the area which was the largest which the great 36-inch refractor of the Lick Observatory could command. It was seen at once that, however powerful the light-grasp of that telescope, it was quite beyond it to give any idea of the structure of so large a body as the Andromeda nebula, when considered as a whole.

But there are other objects in the heavens of far vaster area than the Andromeda nebula, and to deal with these in their full extent requires a wider field than any telescope can cover; they must be observed directly with the unassisted eye.

There are, then, definite branches of astronomy in which the telescope is not only unnecessary, but, more than that, it is a hindrance. Apart, however, from this, it is well to remember that the science was pursued with great success for some thousands of years before ever the telescope was even conceived. The length of the year, the obliquity of the ecliptic, the fact and amount of precession, the chief lunar inequalities, the inclinations of the planetary orbits, and their relative dimensions were all determined by direct eye observation, and with a really remarkable approximation to the truth. Indeed, in our own day the same feat has been repeated, for, as readers of KNOWLEDGE will remember,* there is still living in Orissa the Hindu astronomer, Chandrasekhara, who, with home-made instruments and without optical assistance, has redetermined the elements of the chief members of the solar system with a most astonishing accuracy. Work of this kind may not indeed "increase the sum of human knowledge," for it is to repeat with very small and imperfect means what is being done with the most perfect appliances in the great public observatories of the world. But it is far from being waste time and effort on that account. As a training in keenness of perception and in habits of order and accuracy in observation it will be of the utmost service. It is not every man who climbs the ropes of the gymnasium who expects or wishes to become a sailor, and so to turn the skill he acquires to direct service in exactly the same line; but

---

* See KNOWLEDGE for November, 1899, p. 257.

the strengthening of his muscles and the increase in
agility are solid gains to him none the less.

"What was done in the olden times can be done in the present
day, and I wish to prominently direct the attention of beginners
to the fact that by the employment of quite simple apparatus they
may make observations which will bring home to them in a way
which mere reading can never do, a knowledge of many astronomical
phenomena which they will find to be, not only of immediate
interest, but of great value to them in their further studies.

"What I wish to urge, therefore, is, that those commencing the
study of astronomy should not be content with reading only, but
should work in the open air, faithfully and systematically recording
their observations, however elementary these may be. I lay great
stress on this latter point, because unrecorded observations have,
as a rule, little educational value. The mere fact of describing in
writing any observation, however simple, which has been made is
of immense assistance in securing completeness and accuracy. Of
course, the country offers greater facilities than towns do for this
out-of-door work, but there are few towns where access cannot be
had to some convenient site giving a fairly clear horizon and
sufficiently free from traffic to allow of star maps being referred to
without serious inconvenience. Naturally the beginner's first en-
deavour will be to identify the brightest stars and trace out approximately the confines of the various constellations. Continuing this
study he will gradually acquire a knowledge of the paths followed by
the stars in their courses from rising to setting, and obtain a clear idea
of the position of the apparent axis of this motion. As time goes
on, he will further notice that the constellations he has identified
set earlier and earlier each evening, and that other constellations
previously unseen will come into view on the eastern horizon.
Further, he will notice that the path followed by the moon in her
course through the sky not only differs at different parts of a
lunation, but varies for any given part of a lunation at different
seasons of the year. As his knowledge of the sky progresses, he
will be able to identify any bright planets which may be visible,
and to observe their changes of position with regard to the adjacent

stars, changes which he will do well to note in his sketch-book for future reference and consideration. Now, the beginner who has learned these elementary facts by actual observation of the sky, and has subsequently by the aid of his text-books mastered the reasons for what he has observed, will have made a very fair start in the study of astronomy; and he will, I venture to think, have acquired a far keener interest in the motions of the heavenly bodies than he would have possessed if he had confined his attention solely to books, or if his open-air observations had not been of a systematic character. He will also find that by the aid of some very simple home-made instruments, such as a cross-staff, a rude form of transit instrument, and other similar appliances, he will be able to make observations which serve to still more impress upon his mind the facts he has been learning. Of course, such observations must be crude and wanting in accuracy, but they will, nevertheless, be found to serve a very useful educational purpose."*

It is therefore possible to become a real astronomical observer without a telescope and without any outlay except that necessary to procure a good star atlas. And though it may appear a useless labour thus to traverse for oneself the steps by which the early astronomers attained a knowledge of the universe, yet the value of the training involved will be immense, and the delight to be derived from personally watching in progress the majestic movement of the heavens, the sublimest machine in creation, will soon be felt to be enthralling.

But however great the interest that may be taken in work of the kind just described, the observer will

---

* *Journal of the British Astronomical Association*, Vol. X., No. 1, p. 12.

be sure, ere long, to desire to do something which shall be of value for its own sake, as well as for its secondary effect as training. And, as has been already intimated, there are certain fields, by no means too fully cultivated, which are full of interest, and for which no giant telescopes are required; indeed, in these domains, the unaided eye is the ideal instrument.

First of all, there is the observation of Meteors. The meteors of November attracted a great deal of popular interest in the closing years of last century. Articles and letters in all the newspapers of the land excited general expectation to the utmost. Everyone was anxious to see a display of natural fireworks, exhibited without charge, and which would utterly outdo any efforts of human pyrotechny. It is perhaps no loss to science that the expectation was doomed to disappointment. But though everyone was eager to be a spectator at a magnificent display, there are very few indeed who have cared to become serious observers of meteors. Yet the work is of great interest and value, if systematically carried out; and the work of a single observer, Mr. W. F. Denning, has supplied us to-day with the most perplexing problem that still remains without solution of all astronomy; the problem of " stationary " or " long enduring radiants."

Next, comes the study of the Milky Way. Here

again no telescope is required. A clear sky, keen sight, and great patience are the requisites. And this field is also one which scarcely any observer has taken up. When we have mentioned Heis, Boeddicker, Easton, and Wesley, we have almost exhausted the roll of explorers of the Galaxy. Yet night after night its mysterious convolutions are drawn out athwart the sky, the ring which encloses our universe, the true Mitgard snake that encircles the entire world. Only to the most constant and patient scrutiny will it give up its secrets; yet how large a proportion of the mystery of our Cosmos is involved in an understanding of its structure, who can tell?

Thirdly, there is the Zodiacal Light. We in these high northern latitudes are not well placed for watching it; but it can be seen from time to time, and a thorough use of the opportunities that do come will go far to compensate for our less favourable position. And it is worth mentioning, in this connexion, that the Gegenschein, the faint counterglow to the sun, more difficult and elusive than the Zodiacal Light proper, was independently discovered by an Englishman, and not a dweller in Southern England at that, by Mr. Backhouse of Sunderland. In the Zodiacal Light, and the Gegenschein, we have again objects of the greatest interest and mystery, which are quite unfitted for telescopic examination, are truly naked-eye objects, and which to this day have never been sufficiently observed.

Fourthly, there are Auroræ. At the minimum period of the sunspot cycle there is no reason to expect any immediate recurrence of these beautiful phenomena. But careful training in the knowledge of the constellations and in the three branches of work just mentioned will be the best possible preparation for properly observing Auroræ when they set in again. And this is most important. After a great display it is very easy to collect a number of most vivid and picturesque descriptions, but really useful and scientific accounts are apt to be sadly wanting.

Fifthly, at rare intervals of time we are favoured by the visit of Comets bright enough to be easily seen by the naked eye, and which indeed are sometimes of great magnificence. The determination of the positions of such an object, day by day, or the examination into the minute details of its structure lie beyond the powers of the astronomer without a telescope. But some work still remains for him to do; he can trace out the faint contour of the tail amongst the stars night after night, and from its apparent form can supply some material for deducing the composition of the body.

Sixthly, the rare event of a Total Eclipse of the Sun offers several opportunities for work without the telescope. Drawings of the corona have still a certain use-fulness. Observations of the "shadow bands," of the general illumination as compared with the illumination during twilight; of the stars and planets visible during totality; and of the changes in the coloration of sky and sea and land, all have a certain interest and value.

Of these six branches of astronomy the first four are essentially for the naked eye. There are three others—namely, those of Variable Stars, of the search for New Stars, and of the study of Star Colours—in which an appreciable amount of work may be done without any telescope at all, but in which the possession of a good opera-glass greatly adds to the number of objects which are brought within the observer's reach and to his power of dealing with them. It is my intention, therefore, in this book not to limit myself entirely to work which can be done without any optical aid at all, but to include in "Astronomy without a Telescope," observations for which a good field-glass will suffice.

My programme, therefore, may be divided into four parts. First, the study of the configuration of the con-stellations, so that the principal stars may be easily recognised. And in this connection it may be permissible to touch lightly on the origin and meaning of the con-stellation figures, and of the names of particular stars, which have come down to us from remote antiquity.

For this is also in its measure and degree, "Astronomy without a Telescope." Second, simple observations with the naked eye of the chief apparent movements of the heavens, which may prove serviceable for training in the habits of astronomical work. Third, observations for which the unassisted eye is the proper instrument, that is to say, of Meteors, the Galaxy, the Zodiacal Light, and Auroræ; or for which it may have a limited efficiency in some directions, as with Comets and Total Eclipses. Fourth, observations for which the naked eye is able to accomplish some little work, but in which its efficiency is greatly increased by the help of an opera-glass, as of Variable Stars and of Star Colours.

SECTION I.

CONSTELLATION STUDIES.

SECTION I.

CONSTELLATION STUDIES.

CHAPTER I.

ORIGIN OF THE CONSTELLATIONS.

THE workman is nothing without his tools. For the astronomer in general these are his telescopes; his transit circle; his equatorials. But the fathers of the science had none of these, and they supplied the want by making themselves thoroughly acquainted with the groupings of the stars. The naked-eye astronomer of to-day is compelled to follow their example. The stars are his reference points and he must know them thoroughly; he cannot know them too well, and the more complete and exact his acquaintance with them, the better he is equipped for his work. It is by the stars that he marks the beginning and ending of a meteor's flight; by the stars he lays down the windings and channels of the Milky Way, or the soft contours of the Zodiacal Light. I have felt it, therefore, necessary to preface my notes on the various departments of "Astronomy without a Telescope" by a seriés of "Constellation Studies"; an introduction of the student,

I would hope, to that fuller, more intimate acquaintance with the stellar groupings which continued and careful star-study will soon give to him.

When, where, or why, the constellations were designed and their names given .th^e_m, are questions which have received much attention but which remain without a complete solution. The sources from which light can come on these questions may be divided under four chief heads. First, folk-lore, or oral tradition. This is a rapidly-vanishing factor, and, on that account, it is the more to be desired that those who are brought into contact with the isolated peoples in the corners of the earth should lose no opportunity of trying to find out what these have noticed about the stars, what special groups they recognise, what names they have given them, and what traditions they have preserved about them. Next, what may be called documentary evidence ; allusions in classical writers, and in the astronomical records of India and China. Thirdly, what we may term—to use a popular and convenient, though somewhat inappropriate expression—the " Assyriological " source ; the evidence of monuments and tablets recently discovered in the valley of the Euphrates. This source promises to be the most fruitful and significant, reaching back into a great antiquity, though it has come into our hands but lately. Mr. Robert Brown, Junior, in particular, has followed up this subject for many years with the greatest industry, and has traced back many of the constellations beyond the Greeks from whom we received them, and the Semitic Babylonians, to whom the Greeks were in their turn indebted for them, to those Turanian peoples who inhabited Mesopotamia before the Semitic invasion, the Akkadians and Sumerians. Lastly, there is the evidence of the constellation groups themselves.

This internal evidence is necessarily very limited in its character, yet so far as it goes, it is the most important and unmistakable of all; and is especially valuable when it can be applied as a check to assertions or theories based upon external records of either of the three foregoing categories. To follow up any of these researches is also astronomy—" astronomy without a telescope "— although it is not the astronomy of observation. The chief interest of the evidence afforded by the last-named source is the light which it affords as to the time and place when the constellations were devised, and the reasons it supplies for concluding that they were designed for the most part on a deliberate plan.

It is sufficiently well known to everyone that 48 of the constellations come down from extremely ancient times. The places of the principal stars are given in " Almagest," Ptolemy's great Catalogue, date about 137 A.D. This was a revision of the Catalogue of Hipparchus, date about 140 B.C. There is also a very full description of the constellations in the poem of Aratus of Soli, the " Phenomena," date about 280 B.C., which was a versification of an astronomical work of Eudoxus, date about 370 B.C.

The 48 constellations, therefore, have a clearly recorded history of about 2300 years. But their origin can, by inference, be pushed back to a far earlier date. The poem of Aratus contains clear internal proof that it was not based upon actual observations made in Greece by either Aratus or Eudoxus, but upon a description of the heavens made quite 1500 years earlier. This has been inferred from the references to the places of the equator and tropical circles, and of the rising and setting of stars.

But their origin can be pushed back further still. If

the old 48 constellations are plotted down on a globe, it is at once perceived that they leave a large portion of the sky uncovered. The reason for this is obvious. The stars not included in the 48 constellations did not rise above the horizon of.the place where their designers lived, and, therefore, they did not see them, and could not include them in their scheme. The centre of this void space must have been the celestial S. pole of that date, and its radius gives, approximately, the latitude of the place. This was, roughly speaking, N. lat. 38°, and the date not very far short of 3000 B.C.; say 2800 B.C. For precession has carried round the "first point of Aries," as we now call it, the point where the sun is at the spring equinox, some 66°, corresponding to 4730 years, since the celestial S. pole was in the centre of that unmapped space.

The longitude of the place of origin is not directly indicated, but the presence of the lion and the bear amongst the stellar forms, and the absence of the elephant, the camel, the tiger and the crocodile seem to exclude India towards the east and Europe towards the west, and taken with the indication of latitude already found, confine the search to Asia Minor and Armenia. The internal evidence, therefore, fully con- firms Mr. Robert Brown, when by an entirely inde- pendent method he tracks the constellations back to the head of the Euphrates valley.

The date derived above brings out a number of interesting relations. The constellation figures were all arranged so as to be either upright when on the meridian, or else recumbent; they were not inclined to it. Then the 12 signs of the Zodiac were symmetrically divided by the colures. The spring equinox was in the middle of the Bull, the autumnal in the middle of the Scorpion, the summer solstice in the middle of the Lion, and the winter in the middle of the Water-carrier. The irregularity in the size of these constellations rendered a symmetrical division impossible in later times.

This is an important deduction, for it clears away a number of theories which have been started at different times to account for the origin of the constellation figures, and which assume that Aries, the Ram, was the original equinoctial sign. Even apart from tradition —and tradition is on this point clear and conclusive— it is certain from the position of the constellations themselves that they were mapped long before the spring equinox entered Aries.

Further, so far as the figures themselves allow of the indication, that is in nine cases out of the twelve, all the ascending signs of the Zodiac faced towards the east, all the descending towards the west. This arrangement can scarcely be fortuitous, and is but one of many considerations which afford evidence that the old 48 constellations in general are not the result of chance; one man in one country making a constellation at one time, and another man in another country making another constellation at another time, and so on, but that they form substantially a single document, arranged by one man or set of men, in a comparatively short space of time. Relatively speaking, they were mapped out at one time and in one place.

The people who mapped out these constellations were no savages; they were real astronomers, however elementary their astronomy was. This is clear from two circumstances. First they mapped out the apparent course of the sun during the year by the 12 signs of the Zodiac; by no means a simple thing to do. It might have easily been done indeed, indirectly by means of the moon's mean path, but it is clear that it was done directly and independently of the moon; for the " mansions of the moon," 27 or 28 in number, belong to a different system of astronomy. There is no doubt of the supreme importance of the ecliptic in the astronomy of these unknown ancient people, for the 12 signs by which they marked it, have come down to us by numerous different and independent paths. Egypt, Arabia, Babylon, Persia, and India alike show us the same 12 designs; the minute differences between them only accentuating the reality of their common origin.

A further note is supplied by the care with which the pole of the ecliptic is marked by the folds of the Dragon. The ecliptic is the path of the sun, the lord of light. The pole of the ecliptic is the point furthest from his illuminated course. Here there is fixed the fabled creature which symbolises darkness, cold, eclipse, and night; and with these, evil and death.

To determine the positions of the ecliptic and of its pole was a task far beyond the power of a mere savage. It required no mean amount of skill, observation, and thought. The pole of the equator is easily identified; and the equator can with very little trouble be made out. But these were left unmarked. Had they been precisely marked the date could have been fixed within a few years instead of within a few centuries.

But though the constellations were mapped out by astronomers, they did not consult astronomical convenience in mapping them out. Six obvious canons should have been obeyed had this been the guiding idea.

1. The constellations should have been roughly of equal size. 2. They should have conformed, as far as possible, to the natural configuration of the stars. 3. They should have been compact, not sprawling and

irregular.　4. The designs should have been, as far as possible, distinctive; the same form not being repeated. 5. In no case whatsoever should two constellations have been intermixed.　6. And they should have covered the whole visible sky, leaving no large tracts of space or bright stars unincluded.

All these six canons, except the second, are violated persistently.　The constellations are extravagantly unequal in size.　Ursa Major and Argo occupy spaces on the celestial globe which we might liken to the Russian and British Empires on the terrestrial; whilst the Triangle and Arrow might be paralleled to Sicily and Crete.　Hydra sprawls across 100° of longitude, two-sevenths of the entire circle.　Of the 48 figures, which may be reckoned as increased to 54 by subordinate figures, 38 are the repetitions, more or less frequently, of some 11 forms.　There are ten men, four women, two centaurs, four fishes, four serpentine monsters, two goats, two bears, two crowns, two streams, two dogs, and four birds, including two eagles.　Whilst of intermingled constellations, we have a fine illustration in Ophiuchus, sharing the same stars as Hercules for his head, his foot plunged right into the head of the Scorpion, and his body inextricably mixed up with the coils of the Serpent.

The second canon has had some attention paid to it. One or two constellations have evidently been fitted to the stars which they include, and were possibly even suggested by their natural configuration.　The Northern Crown and the Scorpion are, perhaps, the two best examples, and it has always seemed to me possible that the Great Bear took its name and form from the three strikingly placed pairs of stars which might possibly have suggested the idea of the feet of a great plantigrade animal.　But in general, the natural configuration of the stars gives us no clue whatsoever as to the origin of the constellation figures. Take Pegasus as an example. Here are four stars almost in a square, a striking natural configuration. But one of these stars has been detached to form part of another constellation, Andromeda, and the other three, which form a right-angled triangle, are taken to represent a horse, but not a whole horse; a half horse, winged, and upside down.

Astronomers then designed the constellations, but not purely for astronomical purposes. What purpose had they then? Two are very commonly ascribed to them. The first supposes that the 12 signs of the Zodiac were devised to set forth the seasonal characteristics of the month when the sun was in that particular sign. This explanation is very plausible at first sight, but does not bear examination. First of all, it explains only 12 constellations out of the 48. Next, the constellations of the Zodiac do not correspond to months, but are of most irregular lengths; the Virgin represents nearly six weeks, the Crab not much over a fortnight. Thirdly, there is no country with which we are acquainted the climate of which could be represented by the 12 signs as they stand. And lastly, the theory assumes that Aries was the equinoctial sign, when the Zodiac was devised, whereas Taurus was. The other idea is that the shepherds of Western Asia placed in the sky the fancied representatives of the flocks and herds which formed their wealth, and of the bears and lions from which they had to defend them. This idea is also open to some grave objections. If the starry figures had meant no more than this, and had been freely placed in the sky by anyone who chose, it is impossible to think that the process would have come to so sudden a termination 5000 years ago. Not only was there no effort made to map out the southern circumpolar regions as men became acquainted with them, but large tracts of sky were left unmapped in the northern heavens, and these were left in this state for many generations. Two constellations alone were added to the old 48 within early historic times. The first, Equuleus, is simply a slight expansion and reduplication of Pegasus. The other is Coma Berenices, and obtained a sort of brevet rank through the courtly ingenuity of Conon, the Astronomer Royal to Ptolemy Euergetes. Even here there is some reason to think that he simply gave a new name to an old, an almost forgotten minor asterism. If the constellations had been an evolution in the sense in which this theory supposes, no reason can be suggested why it should have been suddenly arrested 5000 years ago.

Whatever was the history of the constellation symbols it is clear that they were deliberately designed, and designed as a whole. This is proved by the fact that some of the figures are placed in altogether unnatural attitudes, attitudes which have been preserved for thousands of years. Take, for example, Aquarius pouring a stream of water from his ewer, not upon a plant or before his sheep or cattle, but upon a fish which does not swim in the water but eagerly drinks it in. The next sign is more striking still. On the seasonal theory Pisces represents the opening of the fishing season on the break-up of the winter's ice, but the fishes are not represented as netted or hooked by the mouth, but as being tied together by rings slipped over their tails. In other cases the accessories of the figures are clearly significant, *e.g.*, the fish-tail to Capricornus, the lyre round the neck of the eagle Vega, the wings of Virgo, the chain of Andromeda, the chair of Cassiopeia. It is at least probable that there is deliberate design in the circumstance that three figures are truncated, Taurus, Pegasus, and Argo. Then in several cases there' are groups of figures which form something like a connected story; Hercules and the Dragon, Perseus and Andromeda are examples. Even where this is not the case figures of a similar character are gathered together with evident intention; one region of the sky is devoted to the Birds; another to the Fishes and other watery symbols. The Zodiac is clearly intended to form an integral whole, but some of its signs are connected with signs which lie outside, such as Aquarius and the Southern Fish, Ophiuchus and the Scorpion. Of the whole forty-eight constellations, there is scarcely one that is wholly without some sort of connection with neighbouring figures.

The ancient constellations were therefore designed nearly five thousand years ago by a people dwelling somewhere between the Ægean and the Caspian, probably near the head of the valley of the Euphrates, a people which domesticated the sheep, goat, bull, dog and horse, which hunted the bear, lion, and hare, and used the bow and the spear, for all these are represented, and in these connections, in the sky. They had made some progress in astronomy, and were sufficiently organized for the work of constellation making to be carried out on a deliberate plan, and to receive general acceptance.

The constellation figures may not all have been transmitted to the present time without alteration. There is some question about the zodiacal sign of the Balance. Ursa Major and Ursa Minor may not originally have been bears, but rather waggons or chariots, and so also with two or three other groups. But many significant little details seem to show that the constellations, as a whole, have been preserved without important change.

CHAPTER II.

The Great Star Clock in the North.

THE stars wear a very different aspect to the astronomer with a telescope and the astronomer without. The former, deep in his observatory dome, sees but a narrow slice of the sky through the open shutter, and the starry groupings as such have little or no significance for him. If he wishes to bring a star within the field of his instrument, he does not as a rule seek it out first on the sky, and then aim his telescope at it, like a rifle, by its sights. Instead he refers to his catalogue, reads therein the right ascension and declination of the object, turns his instrument until its circles are set to the readings indicated by the catalogue, and then, last of all, moves his dome round until the shutter opening is opposite the object glass. The names of the stars, the constellations in which they are found, have therefore very little significance for him. The important things for him to know are the hour, minute and second to which the one circle must be set; the degree and minute to which the other.

Not so with his brother worker. He stands out under the open heaven; no graduated circles guide his gaze to this star or that. For him, if he will know precisely to what part of the heavens he is directing his attention, it is necessary to be able to recognise the individual stars. In this work differences of brightness and colour are no small help, but by themselves would be perfectly inadequate guides to the recognition of the great majority of the stars. That by which one star can be recognised from another is in most cases its grouping with the rest. The knowledge of such grouping, a perfect and quick recognition of the figures, real or imaginary, which the stars make up amongst themselves, in a word a knowledge of the constellations, is the first essential for the direct observer. It was so from the very beginning. The first astronomers necessarily had no telescopes, and equally of necessity the first great astronomical enterprise was the dividing out of the heavens into constellations, the ascribing certain imaginary figures to particular groups of stars, and the bestowal of names upon individual stars themselves.

The same necessity makes itself felt in every branch of science. Before any progress can be made the objects recognised in that science must be named. Until they are named they are undistinguished and undistinguish-able. So far as we are concerned they remain without properties, one might almost say without existence; once named, a knowledge of their properties and peculiarities begins and a whole new field of research is opened out.

And even without this further knowledge, how great an interest is given to any object by the fact that we know its name. Take some town children out into the country, and set them to gather wild flowers, how instantly they ask their names, and how much their beauty is increased in their sight when those names are taught them. And so to-day we are continually hearing the complaint of Carlyle repeated :

"Why did not somebody teach me the constellations, and make me at home in the starry heavens, which are always overhead and which I don't half know to this day?"

So the work of learning the stars, though it may involve some self-denial, and brings no reward in the shape of "magnificent spectacles," has a charm of its own. The silent watchers from heaven soon become each one a familiar friend, and to any imaginative mind the sense that he is treading the same path as that traversed by the first students of Nature will have a strange charm. With the "Poet of the Breakfast Table" he will feel himself linked to the great minds of the deep unmeasured past.

"I am as old as Egypt to myself;
"Brother to them that squared the Pyramids.
"By the same stars I watch."

However often, therefore, the work of teaching the constellations may have been undertaken, it forms an inseparable portion of my present task.

To us in England, with our high northern latitude, the stars which never set are the first the study of which we should undertake. They are always present, they cover more than one-third of the entire sky visible to us at any moment. They include many conspicuous stars, and form an admirable guide to the constellations beyond the circumpolar region. Constantly revolving round the pole they form, as it were, a magnificent dial plate, marking at the same time the progress both of the night and of the year.

The chief constellation of this region is the Great Bear, the leading stars of which are the Seven, which have won the attention of all races of men in all ages. The

seven stars of the Plough or Charles' Wain (the waggon, that is to say, of the churl or peasant) are known to everyone, and form the inevitable starting point for the study of the constellations. Of these seven stars,—which, at midnight on the first of April are practically overhead, the greater part of the constellation being already on the downward path towards the west,—the two first are Alpha and Beta, the second pair Gamma and Delta, the four making up the body of the plough, whilst Epsilon, Zeta, and Eta form the handle. Delta is distinguished as being much the faintest of the seven, Zeta by its close companion, Alcor, visible to any ordinarily good sight.

Each of the seven stars of the Plough, has, like Alcor, its own special name derived from the Arabic, but they are more generally known by the letters of the Greek alphabet,* and are referred to as Alpha, Beta, Gamma, etc., Ursæ Majoris. The other chief stars of the constellation are very seldom referred to except by the Greek letters, their Arabic names having passed out of use.

Regarding the constellation as the "Great Bear," the four stars in the body of the Plough make the hindquarters of the animal, whilst the handle becomes the bear's tail. The feet of the bear are clearly pointed out by a curious set of three pairs, Iota and Kappa make the first, Lambda and Mu the second, Nu and Xi the

---

* The Greek Alphabet is as follows :—

α	Alpha	η	Eta	ν	Nu	τ	Tau
β	Beta	θ	Theta	ξ	Xi	υ	Upsilon
γ	Gamma	ι	Iota	ο	Omicron	φ	Phi
δ	Delta	κ	Kappa	π	Pi	χ	Chi
ε	Epsilon	λ	Lambda	ρ	Rho	ψ	Psi
ζ	Zeta	μ	Mu	σ	Sigma	ω	Omega

third. These form the great plantigrade feet of the animal, and are " the does' leaps " of the Arabs.

A line drawn from Zeta through Alpha and carried forward the same distance the other side brings us to the fourth magnitude star Omicron at the point of the creature's snout. These stars enable the boundaries of the constellation and the figure which it is supposed to represent to be easily detected in the sky. Beta and Alpha are commonly known as the " Pointers," for as the " poet " sings :

> "Where yonder radiant hosts adorn
>   The northern evening sky,
> Seven stars a splendid glorious train
>   First fix the wandering eye.
> To deck great Ursa's shaggy form
>   Those brilliant orbs combine,
> And where the first and second point
>   There see Polaris shine."

A straight line from Beta through Alpha points very nearly up to the pole of the sky, the distance being just a little greater from Alpha to the pole than from Alpha to Eta, and close to the pole shines the Pole star, a brilliant of the second magnitude, and placed at the end of the tail of the Lesser Bear as Eta is at the tip of that of the Greater.

Starting from Epsilon Ursæ Majoris, the star in the Great Bear's tail nearest the root, and crossing the North Pole, we find on the further side of the Pole, right upon the sparkling background of the Milky Way, here almost at its broadest, five stars in the shape of a W, the principal stars of the constellation Cassiopeia, the " Lady in her Chair." At midnight on the first of April, this group is low down in the north ; the W being, as it were, written in a dropping line from left to right, that is from west to east, as if scrawled by a tired writer. The lettering of the stars is nearly but not quite in the reverse order of writing. Reading from left to right they come Epsilon, Delta, Gamma, Alpha, Beta, the three last named being distinctly brighter than the other two.

Starting from Cassiopeia, and following the Milky Way towards the west, we find a number of stars marking out the spine of the Galaxy, and bending down in an elegant curve to the bright somewhat yellow star in the north-west. The stars in this curve are the principal members of the constellation Perseus, and the bright yellow star at the end of the curve is Capella, Alpha in the constellation Auriga. It is a star impossible to mistake, since close beside it is a very pretty little right-angled triangle of moderately bright stars.

Not far from Capella is a bluish-white star of the second magnitude, Menkalinan, or Beta, in the constellation Auriga, in which Capella is Alpha. On a straight line from Beta Aurigæ through the Pole Star, and rather further from the Pole Star on the other side, is a splendid steel-blue star, Vega, the rival of Capella in brightness, the two being claimants for the premiership of the northern heavens. The five bright stars which wait upon Vega in its immediate neighbourhood, and of which the nearest, Epsilon, is a very close double to keen sight, make up with it the constellation of Lyra, a constellation which lies for the most part outside the circumpolar circle for the latitude of London.

The chief guiding stars, therefore, for the northern heavens, are the well-known Plough, the scarcely less distinctive little W of Cassiopeia on the opposite side of the Pole, and the two great brilliants between them on the right hand and on the left, Capella and Vega. All these are continually visible for Scotland and the

North of England ; for the southern part of our island
Vega is lost for a short time when due north.

To watch these northern constellations as they follow
each other in regular ceaseless procession round the
Pole, is one of the most impressive spectacles to a
mind capable of realizing the actual significance of
what is seen. We are spectators of the movement of
one of Nature's machines, the vastness of the scale of
which and the absolutely perfect smoothness and

ZENITH.

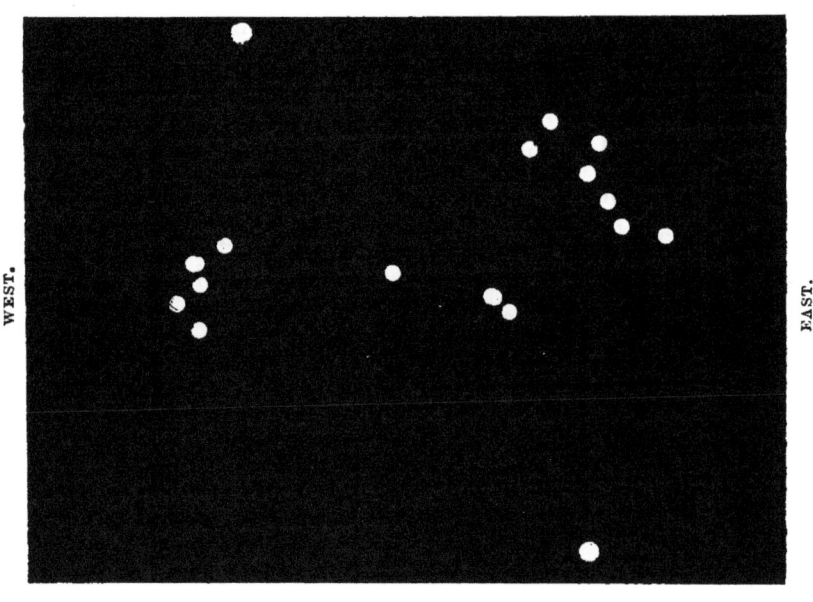

WEST.

EAST.

HORIZON.

FIG. 1.—Looking north at Midnight on JANUARY 13.

regularity of whose working so utterly dwarf the
mightiest work accomplished by man. The sense of
this ceaseless motion and of its perfect regularity sank
deep into the minds of the earliest observers, and had
much to do with the sacred, or at least semi-sacred
character, which attached to the study of astronomy

in those ages. But beyond this, there was the actual practical value of the movements of this great celestial clock. An observer, watching through the hours of a winter's night, will see the Plough low down in the north at first, raise itself little by little towards the east, reaching the zenith about 5 o'clock in the morning, and at daybreak be moving downwards in the

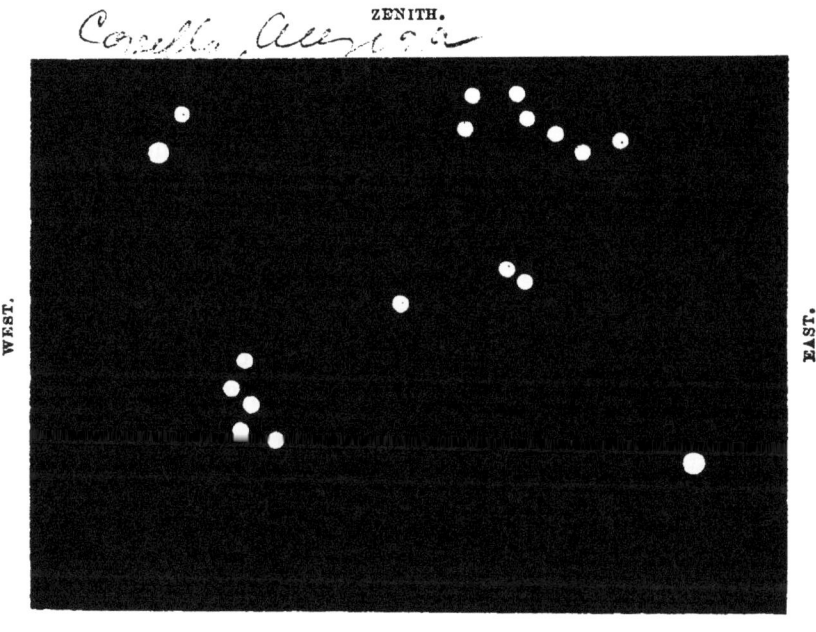

FIG. 2.—Looking north at Midnight on FEBRUARY 13.

west. Since the stars reach any given position in the sky about four minutes earlier on each successive night, the positions of these four constellations not only show how far the night is advanced on any particular occasion, but if observed night after night at the same hour they act as a calendar to mark the progress of the year. The twelve little diagrams here given are intended to show roughly the relative positions of these four constellations and of the Pole Star as seen by an observer in the latitude of London who faces due north. The first diagram shows their aspect as seen at midnight in the middle of January, the next gives the same view for midnight in the middle of February, the third for midnight in the middle of March, and so on. Or we may regard them as representing two-hourly intervals in the course of any given night. Thus for the night of January 1, No. 9 (the September diagram) would represent the appearance of the sky at 5 o'clock in the evening, No. 10 at 7 o'clock, No. 11 at 9 o'clock, No. 12 at 11 o'clock, and No. 1 at 1 o'clock in the morning. Nos. 2 and 3 and 4 would in like manner represent the appearance at 3, 5 and 7 o'clock in the morning respectively, so that eight out of the twelve diagrams would be seen during a single night.

The following table shows the hours during any particular night represented by the different diagrams, and the dates throughout the year which they represent for any given hour of the night :—

DAYS AND HOURS REPRESENTED BY THE DIAGRAMS OF THE NORTHERN STARS.

HOUR.	I.	II.	III.	IV.	V.	VI.	VII.	VIII.	IX.	X.	XI.	XII.	HOUR.
h. m.													h. m.
5 0 P.M.	......	......	......	......	......	......	Oct. 29	Nov. 29	Dec. 29	Jan. 29	......	......	5 0 P.M.
6 0	Mar. 31	......	......	......	......	......	Oct. 14	Nov. 14	Dec. 14	Jan. 13	Feb. 13	......	6 0
7 0	Mar. 15	April 15	May 15	......	......	Aug. 30	Sept. 29	Oct. 29	Nov. 29	Dec. 29	Jan. 29	Feb. 28	7 0
8 0	Feb. 28	Mar. 31	April 30	May 30	June 30	Aug. 14	Sept. 14	Oct. 11	Nov .14	Dec. 14	Jan. 13	Feb. 13	8 0
9 0	Feb. 13	Mar. 15	April 15	May 15	June 14	July 30	Aug. 30	Sept. 29	Oct. 29	Nov. 29	Dec. 29	Jan. 29	9 0
10 0	Jan. 29	Feb. 28	Mar. 31	April 30	May 30	July 15	Aug. 14	Sept. 14	Oct. 14	Nov. 14	Dec. 14	Jan. 13	10 0
11 0	Jan. 13	Feb.	Mar. 15	April 15	May 15	June 30	July 30	Aug. 30	Sept. 29	Oct. 29	Nov. 29	Dec. 29	11 0
Midnight	Dec. 29	Jan. 29	Feb. 28	Mar. 31	April 30	June 15	July 15	Aug. 14	Sept. 14	Oct. 14	Nov. 14	Dec. 14	Midnight
1 0 A.M.	Dec. 14	Jan. 13	Feb. 13	Mar. 15	April 15	May 30	June 30	July 30	Aug. 30	Sept. 29	Oct. 29	Nov. 29	1 0 A.M.
2 0	Nov. 29	Dec. 29	Jan. 29	Feb. 28	Mar. 31	May 15	June 14	July 15	Aug. 14	Sept. 14	Oct. 14	Nov. 14	2 0
3 0	Nov. 14	Dec. 14	Jan. 13	Feb. 13	Mar. 15	April 15	May 30	June 30	July 30	Aug. 30	Sept. 29	Oct. 29	3 0
4 0	Oct. 29	Nov. 29	Dec. 29	Jan. 29	Feb. 28	Mar. 31	May 15	......	......	Aug. 14	Sept. 14	Oct. 14	4 0
5 0	Oct. 14	Nov. 14	Dec. 14	Jan. 13	Feb. 13	Mar. 15	......	......	......	......	Aug. 30	Sept. 29	5 0
6 0	......	......	Nov. 29	Dec. 29	Jan. 29	......	......	......	......	......	......	......	6 0
7 0	......	......	......	......	......	......	......	......	......	......	......	......	7 0

D

ZENITH.

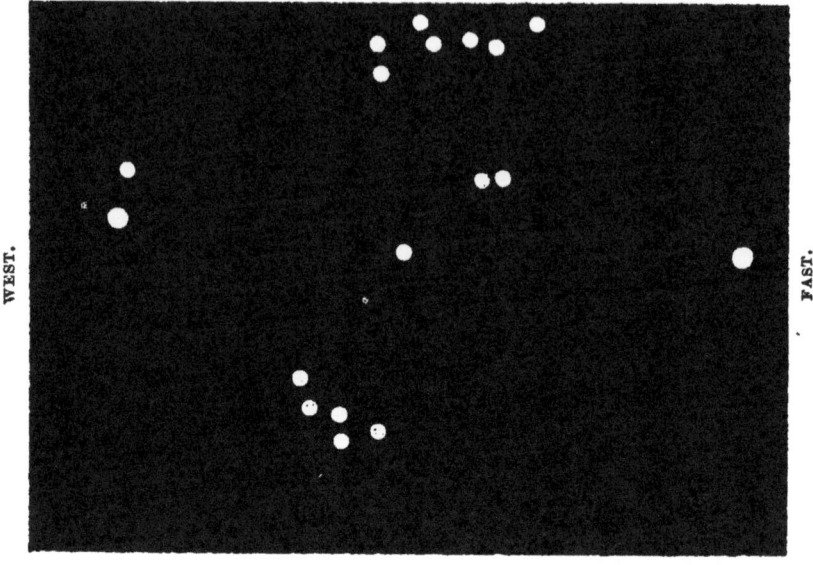

HORIZON.

FIG. 3.—Looking north at Midnight on MARCH 15.

ZENITH.

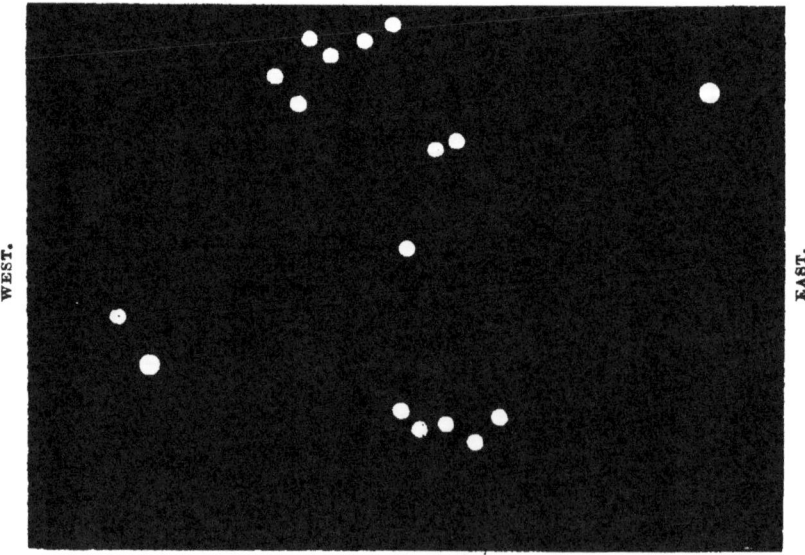

HORIZON.

FIG. 4.—Looking north at Midnight on APRIL 15.

ZENITH.

WEST.

EAST.

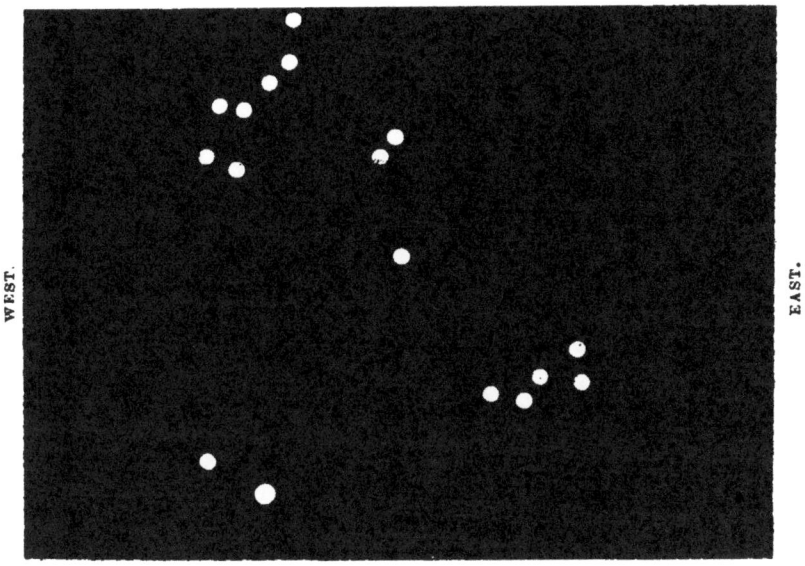

HORIZON.

FIG. 5.—Looking north at Midnight on MAY 15.

ZENITH.

WEST.

EAST.

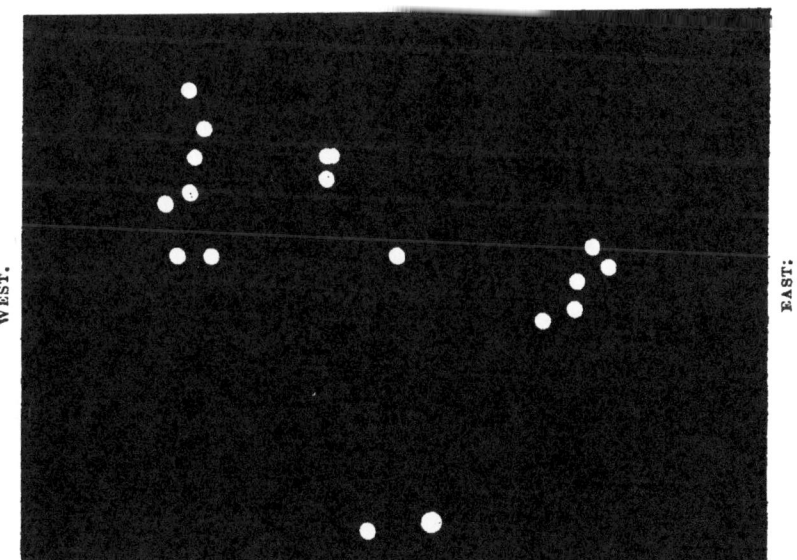

HORIZON.

FIG. 6.—Looking north at Midnight on JUNE 14.

D 2

ZENITH.

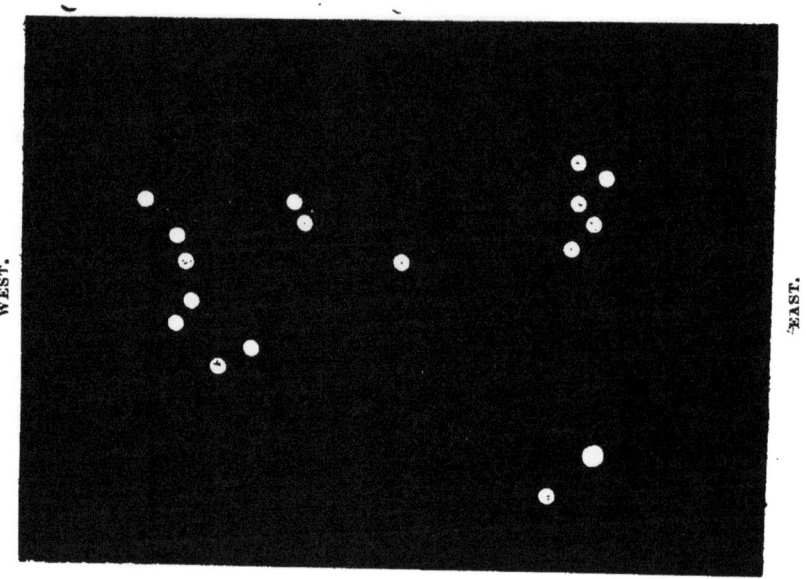

HORIZON.

FIG. 7.—Looking north at Midnight on JULY 15.

N

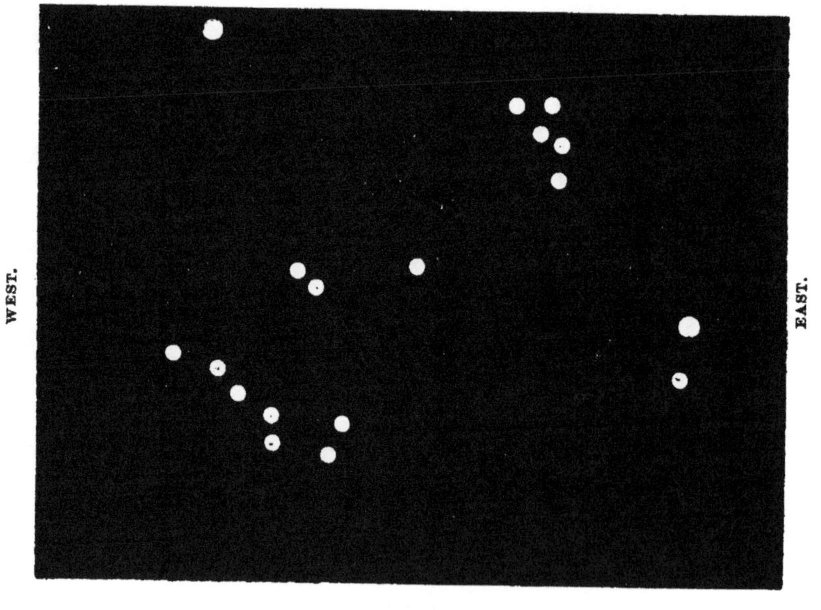

HORIZON.

FIG. 8.—Looking north at Midnight on AUGUST 14.

ZENITH.

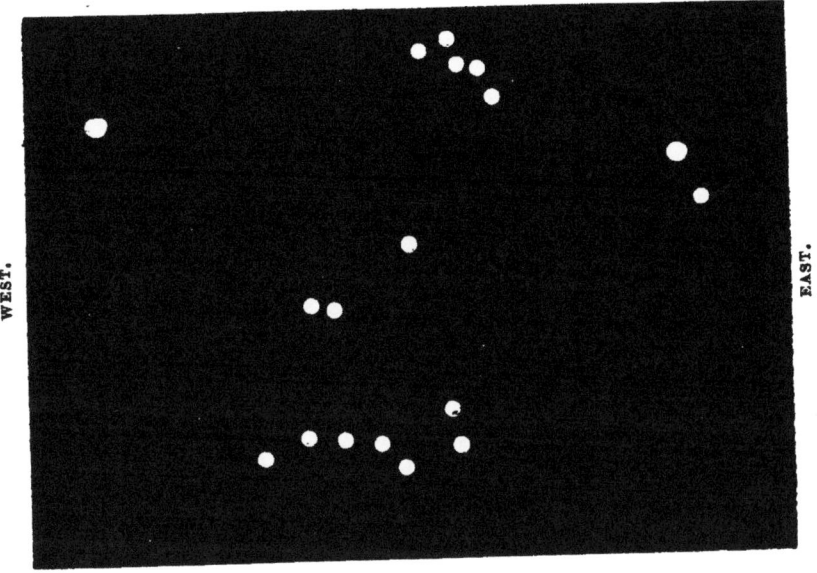

HORIZON.

FIG. 9.—Looking north at Midnight on SEPTEMBER 14.

ZENITH

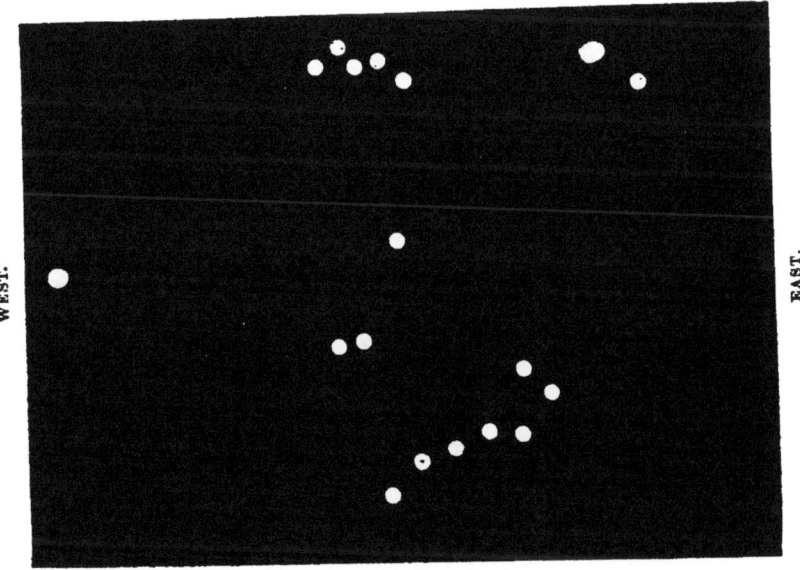

HORIZON.

FIG. 10.—Looking north at Midnight on OCTOBER 14.

ZENITH.

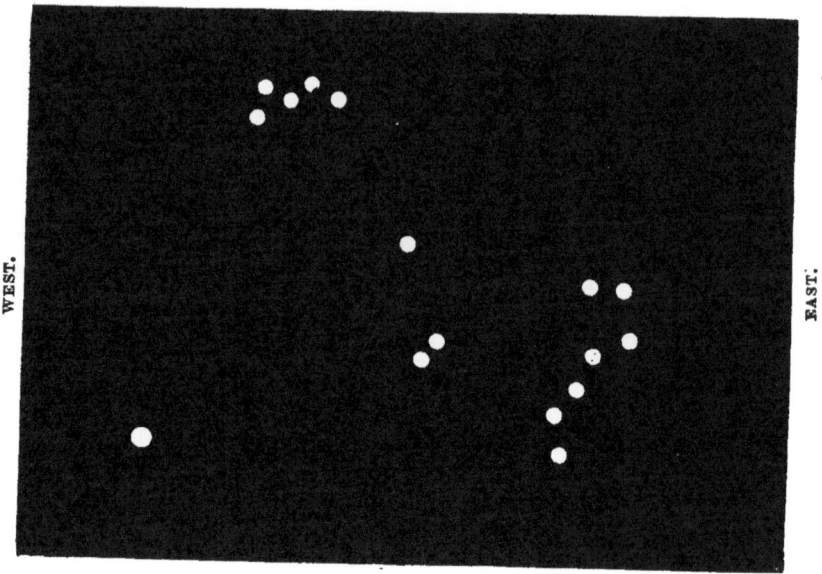

HORIZON.

FIG. 11.—Looking north at Midnight on NOVEMBER 14.

ZENITH.

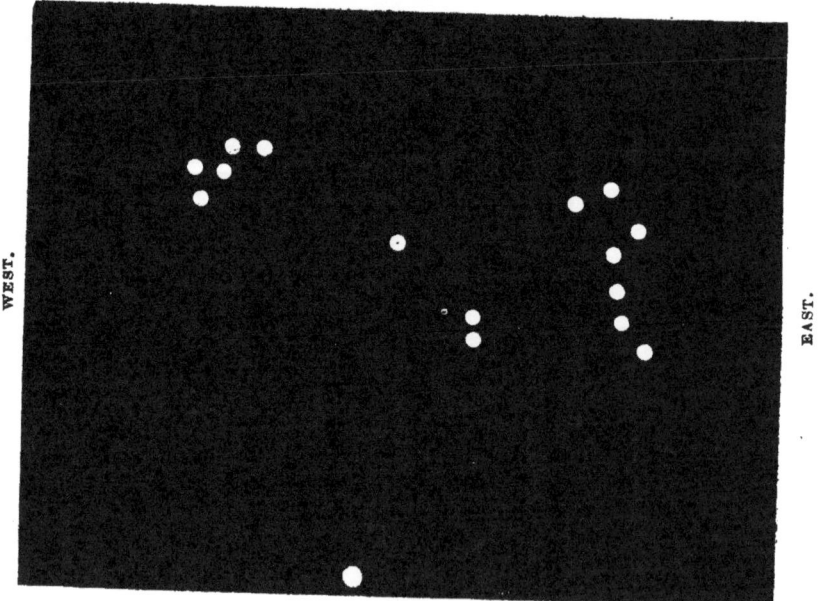

HORIZON.

FIG. 12.—Looking north at Midnight on DECEMBER 14.

In diagram No. 1 Capella is very nearly in the
zenith, Vega right down on the northern horizon,
whilst the Plough is high on the east, and Cassiopeia
is level with the Pole Star on the west of it.   In
No. 4 the last star in the Plough Handle is not seen,

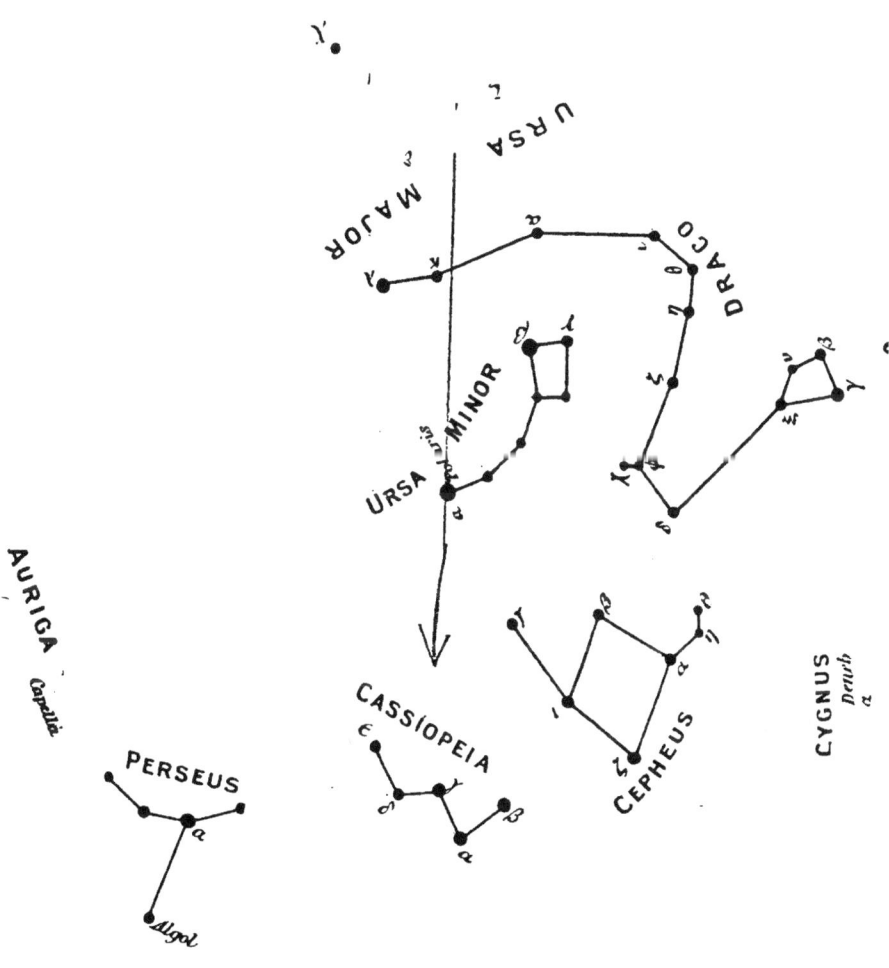

Fɪɢ. 13. — The Chief Circumpolar Stars, midnight, April 1st.

for it passes the meridian a little to the south of the
zenith, and therefore is right over our observer's head,
and, since he is facing northward, somewhat behind
his back. In diagrams 5, 6 and 7 Vega is not seen,
for in all these that star is to the south, and therefore behind the observer's back. In Nos. 11 and 12
Capella and Beta Aurigæ are out of view in the same
manner. In No. 12 the place of Vega just below the
horizon in the north is indicated.

Besides the seven stars of the Plough, the five stars
of Cassiopeia, Vega, Capella, and Beta Aurigæ, three
stars are shown of Ursa Minor. The centre star of
the diagrams is of course the Pole Star, Alpha Ursæ
Minoris; the other two stars shown are the other two
brightest stars of the constellation, Beta and Gamma,
and they are added to indicate the way in which the
Lesser Bear swings round in the sky with the tip of
his tail as pivot.

The accompanying map shows the position of the
circumpolar region with regard to the north horizon at
midnight on the first of April. The figures ranged
round the circumference of the map show the position
of the north point of the horizon for hourly intervals
of the day and night at that time of the year. For other
dates in the year we can find its position nearly enough
by remembering that for every month later in the year
that we take we must also take two hours earlier in the
evening to obtain stars in the same position, or if we
take a single day later in the year then we must choose
our time four minutes earlier.

CHAPTER III.

THE NORTH CIRCUMPOLAR STARS.

THE first task in the study of the constellations is to learn to recognise the Pole Star and those chief jewels which circle round it, and which mark out with sufficient nearness the four quadrants – Cassiopeia, Capella, the Plough, and Vega. When familiar with these and with their changing positions relative to the zenith and horizon, as they swing round the celestial pole during the successive hours of the night and months of the year, it is easy to go on to the study of the lesser details of the great circumpolar region.

The constellations that are either wholly included in this region, or very nearly so, are eight in number, five of them ancient; the other three, Camelopardus, Lynx, and Lacerta, were added by Hevelius about 1690, but include no bright stars, and possess no easily recognised features. The five ancient constellations, therefore, claim our first attention.

URSA MAJOR.

There is no place for hesitation as to which of these constellations we should begin with.

"He who would scan the figured skies,
 Its brightest gems to tell,
 Must first direct his mind's eye north,
 And learn the Bear's stars well";

the seven stars so well known to our own peasantry as the " Plough " or " Charles' Wain." Wherever men have taken any notice of the stars at all these seven have been recognised as a natural group, and in earlier ages, being then much nearer to the Pole than now, they were amongst the stars always visible, not only to dwellers in such northern latitudes as our own, but as far south as the tropic of Cancer. It is easy to see how the names of " Plough " or " Wagon " for these seven stars have arisen; their natural configuration has suggested them. The three stars below, as we look at the constellation at midnight in the autumn of the year, suggest just the kind of curve of a plough handle; and the four above in a rough rectangle, present the plough-share. Or the four stars above may be considered the four wheels of the rude wagon of which the three below represent the heads of the three horses. " Chariot " or " Wagon " the seven stars have been not only in Northern Europe in our own time but in ancient Greece, and still more ancient Babylonia. Aratus writes of the Pole : —

" Two Bears
Called Wains moved round it either in her place."

And Homer says that on the shield of Achilles were

"All those stars with which the brows of ample heaven are crowned,
Orion, all the Pleiades, and those seven Atlas got—
The close-beamed Hyades, the Bear surnamed the Chariot."

But how the constellation got the name of the Bear is far harder to explain. The Sanskrit name " Riksha " signifies both " Bear " and " Star," that is, " bright " or " shining " one, and the latter word—very justly applicable to the seven stars, as being pre-eminently the stars, the shining ones of the northern sky—may perhaps have been punningly represented by the figure

of a bear. But this assumes that the title is Aryan
in its origin, which is indeed far from certain. In
default of a better theory I am myself inclined to think
that the three striking pairs of stars below the Plough
suggested the feet of a great plantigrade animal. The
lesser Bear no doubt obtained its name from the greater,
since its principal stars are a distorted and fainter copy
of the seven brilliants of its near neighbour. Classical
tradition, according to Aratus, held that they were
transferred to heaven as a reward for hiding Zeus in
Crete, from his cannibal father Kronos, or else the
Great Bear is Callisto, one of Zeus's many loves, and
Arcas, the Lesser Bear, her son. The seven great stars
of the Plough are now known by the first seven letters
of the Greek alphabet, proceeding in order from the
front of the ploughshare back to the handle. The
names which they popularly bear at the present day
are as follows:—Alpha is Dubhe, that is, the "bear";
Beta, Merak the "loin"; Gamma is Phecda the
"thigh"; Delta, the faintest of the seven, is Megrez,
"the root of the tail"; Epsilon is usually called
Alioth; but whether this name has much authority is
not clear; Zeta is Mizar, a "girdle" or "waistcloth"—
but this is a comparatively modern appellation; Eta,
the star at the tip of the tail, has the most interesting
name of all, since it is called Alkaid or Benetnasch;
the two names together meaning the "chief of the
daughters of the Bier." It will be remembered that
in Job, Chap. xxxviii., the patriarch is asked, "Canst
thou guide Arcturus with his sons?" "Arcturus" being
the erroneous rendering adopted in the A.V. for "*Aish*"
the "Bier" or the "Assembly." This star preserves to
us, therefore, almost unchanged the name which the

constellation bore at the time when the great drama of Job was written.

By far the most interesting object in the whole constellation to the "astronomer without a telescope" is Mizar with its near companion Alcor, 80 in Flamsteed's enumeration. Mizar is in every way the first of the double stars. Alcor forms with it a double to the eye; it has a much closer bright companion which rendered it the first double star to be detected in the telescope, it was the first double star to be photographed, and it was the first case in which the spectroscope showed that the principal star, which appears to us even with the most powerful telescope as single, is really in itself double. Epsilon Ursæ Majoris marks very nearly the place of the radiant point of a shower of Ursid meteors, the date of which is the 30th of November. For those astronomers who add the opera-glass to naked-eye work, the three stars of the plough handle and their immediate neighbourhood offer many interesting fields. So, too, the feet of the Bear, the three pairs of stars to which we have already alluded, are also worth studying with this amount of optical aid. The fore foot is composed of Iota and Kappa, Lambda and Mu mark the next, Nu and Xi the last.

URSA MINOR.

Merak and Dubhe are, as is well known, commonly called the "Pointers," inasmuch as the straight line drawn through them leads us very nearly to the Pole Star, which is about the same distance from Dubhe as Dubhe is from Alkaid. The name given it, from its nearness to the Pole, Polaris, is so universally applied to it, nowadays, that there is little need to notice the many Arabic names which it has borne. An opera-glass, however,

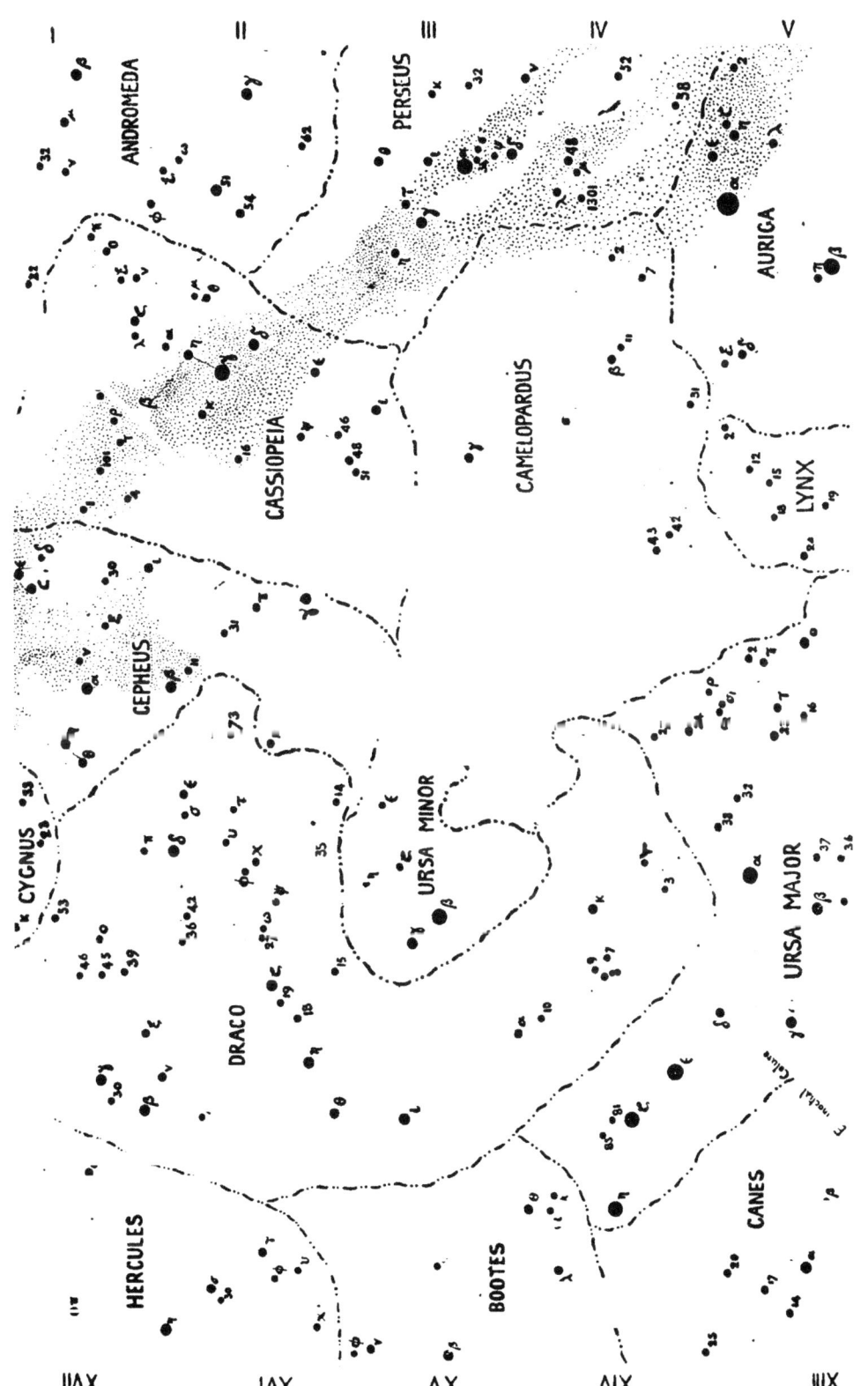

shows us other fainter stars yet nearer the Pole, of which the chief is Lambda Ursæ Minoris, just on the limit of unassisted vision, whilst Groombridge 1119 is fainter still. Three other stars, visible to the naked eye, may be mentioned as within $3\frac{1}{2}°$ of the Pole ; Cephei 51, which lies within the boundaries of Ursa Minor as the constellation is usually drawn nowadays, though it bears the foregoing designation, which it owes to Hevelius; Delta Ursæ Minoris ; and Bradley 3147. This little group of stars, though not attractive to the sight, is of the utmost importance astronomically, since its components enable the professional astronomer to test the accuracy with which his transit instrument points to the north at intervals of about two hours.

Beta and Gamma are the only other conspicuous stars of the constellation. Beta was once the Pole Star, or at any rate divided the title rôle with Kappa Draconis, and hence bears the name Kochab, "The Star," that is to say, the Northern or Pole Star. Gamma is a wide double star to the eye, Al Farkadain, the "calves," usually written Pherkad on our globes.

The constellation is the seat of several of the minor radiants. In September, there is one from the neighbourhood of 51 Cephei ; in April, there is another from near Gamma Ursæ Minoris.

DRACO.

Between the two Bears, and almost encircling the Lesser, is a long winding stream of stars, making up the constellation of the Dragon. It is certainly one of the most ancient of all, and is believed by many to be the crooked serpent of Job xxvi. 13. This is the Serpent of the following lines from "Aratus" :*

---

* Brown's "Aratus," p. 16.

> "Between these two, like to a river's branch,
> A mighty prodigy, the Serpent twines
> Its bendings vast around; on either side
> His coil they move and shun the dark blue sea.
> But o'er the one his lengthy tail is stretched,
> The other's wrapped in coil."

Alpha Draconis, sometimes called Thuban or Rastaban, lies midway between Zeta Ursæ Majoris and Gamma Ursæ Minoris. Alpha Draconis was the original Pole Star of the heavens when the constellations were mapped out, a pre-eminence it must have held for over 2000 years. Within this constellation also is the pole of the ecliptic, almost in the centre of the great loop made by the Dragon's folds.

At midnight on New Year's Day, the Dragon's head reaches down almost to the northern horizon, two bright stars, Gamma and Beta, marking the top of its head. Of these the more westerly is of a rich orange tint, and is the zenith star of Greenwich, and as such was specially observed by Flamsteed, Bradley, and Airy, the second of whom made his discovery of the aberration of light in connection with it. Three stars, Xi, Nu, and Mu, make up the jaw, Mu being at the snout. Nu, the faintest of the three, is an opera-glass double.

CASSIOPEIA AND CEPHEUS.

From Epsilon in the Great Bear a line through Polaris leads us to a small constellation, yet one of the most easily recognisable in the sky, Cassiopeia, the "Lady on her Throne," her principal stars, five in number, suggesting a W freely scrawled. The only stars of the five which in modern days are often referred to by their Arabic names are Alpha, Schedar or "Breast"; and

Beta, Caph or " Hand," or possibly in some allusion to her husband Cepheus who stands by her side.   The latter forms a larger but much less conspicuous constellation, lying between Cassiopeia and the Dragon, its four chief middle stars forming a lozenge ; the point of the lozenge most remote from Cassiopeia is Alpha, Alderamin, the "right arm," the only one commonly referred to now by its Arabic name.   Delta is one of the most interesting of the naked-eye short period variables, and an opera-glass double.

Cepheus and Cassiopeia are especially interesting since they with three more southern constellations make up a recognised and unmistakable story pictured in the sky ; a clear proof that the work of original constellation making was deliberate and not haphazard, and that the legends there represented were in existence before the star groups were made.   Brown argues justly that Cepheus is manifestly a non-Hellenic sovereign.   He is indeed often spoken of as Ethiopian, but the Ethiopia there meant is not Nubia or Abyssinia, but the Euphratean " Cush."   Hence there is no justification for those too precise artists and poets who have represented poor persecuted Andromeda as a sable beauty, " black" if " comely."

Cassiopeia is a constellation that well repays opera-glass scrutiny, and it also furnishes to the naked-eye astronomer several important meteor radiants, of which one from near Delta deserves especial notice.   It is also a particularly favourable neighbourhood in which to commence the study of the Milky Way, since the constellation passes through our English zenith.   It is most famous historically from the appearance of the celebrated Nova of 1572—the "Pilgrim Star"—which formed very nearly the fourth point of the square, or rather rhombus, of which Alpha, Beta, and Gamma mark the other three points.

Camelopardus is a great straggling constellation, all the stars of which are faint, which lies between Ursa Major and Cassiopeia, and stretches upwards almost to the Pole.

Another group is Lacerta, the Lizard, a few faint stars gathered together by Hevelius to form a constellation fitted in between Cepheus, Cygnus, Pegasus and Andromeda. Except that it marks the radiant point of the Lacertids, a meteor shower, active in August and September, there is nothing to distinguish the group. Apparently at one time the foreleg of Pegasus crossed this region, since Pi Cygni, which lies beyond it from Pegasus, bears the name Azelfafage, the "hoof of the horse."

CHAPTER IV.

THE STARS OF SPRING.

THE four groups of bright stars surrounding the Pole, namely Cassiopeia, Capella, the Plough, and Vega, not only serve as hour marks on the dial plate of the sky, but are most useful as guides to the other leading constellations. The Plough occupies the zenith at midnight during the month of March and the first fortnight of April. It may therefore be usefully taken as furnishing a guide to the stars in view during spring. Vega is overhead at midnight at the end of June and thus rules over the stars of summer. Cassiopeia occupies the crown of heaven during the first fortnight of October, and is thus the queen of autumn. Whilst Capella and his companion Menkalinan ride at their highest during the nights of the middle of December, and preside over the stars of winter.

LEO.

The nights of spring bring to the meridian the most famous of all the constellations of the Zodiac; the constellation, that is to say, of the Lion. Its primacy is beyond question due to the fact that the place of the sun at the summer solstice was in this constellation at the time when they were first devised, and no doubt its brightest star derived its name, Regulus, or "little king,"' from being the chief star of the paramount sign. Both names are traditional in many different countries; the constellation is the Latin Leo, the Hellenic Λεων, the Persian Shir, the Hebrew Aryeh, and the Babylonian Aru, all alike meaning " Lion "; whilst our present name for the star is the variant, proposed by Copernicus, for the older Latin Rex. Ptolemy calls it βασιλισκός, the Arabs give it Malikiyy, the " kingly " star, and the cuneiform inscriptions of the Euphratean

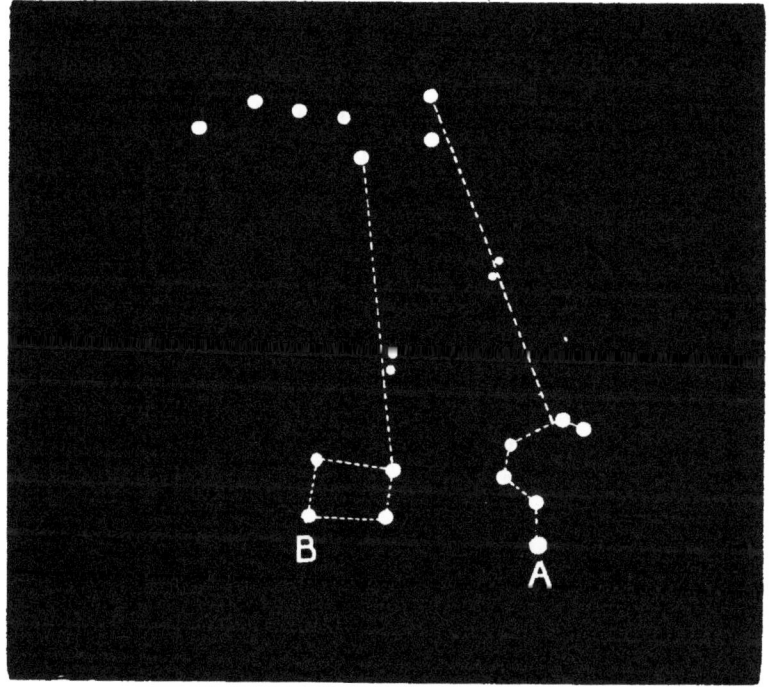

A, Regulus.   B, Denebola.

FIG. 14.—The Lion and the Plough.

valley refer to it as the " star of the king," whilst in ancient Persia it was the chief of the four " royal stars." It is its place, however, and not its brilliance, which has gained for Regulus this distinction, for almost all the first magnitude stars are its superiors in brightness.

The constellation of the Lion is very easily found when the Great Bear is known.

> "'Neath her hind feet, as rushing on his prey
> The lordly Lion greets the lord of day."

The Great Bear at this season at midnight is at its greatest elevation, and below it towards the south, we find the Lion. The stars in it are formed into two principal groups; the Sickle, six bright stars marking the animal's head and breast, whilst a Rectangle indicates its hinderquarters. A line from Alpha in the Great Bear through the third foot, that marked by Lambda and Mu, and prolonged beyond the foot to an equal distance, brings us to the centre of the blade of the Sickle, whilst another line from Gamma through the fourth foot leads to the Rectangle.

The stars of the Sickle, beginning with the most westerly, run in the following order, Epsilon Mu, Zeta, Gamma, Eta, Alpha. In the very centre of the trapezium made by the first four of these, is the place of the radiant of the celebrated Leonid shower of meteors, the showers which gave us such splendid displays in 1833 and 1866, and to which in truth we owe our knowledge of meteoric astronomy, since they first drew real scientific attention to the subject of meteors and afforded the means of solving many of the problems which they present.

The fourth star Gamma is but little inferior to Regulus in brightness, and in a telescope it is an extremely interesting and beautiful double star. It bears the Arabic name Algieba, meaning "Forehead," though it is actually situated on the Lion's breast. It forms a fine contrast in colour to Regulus, being distinctly deep

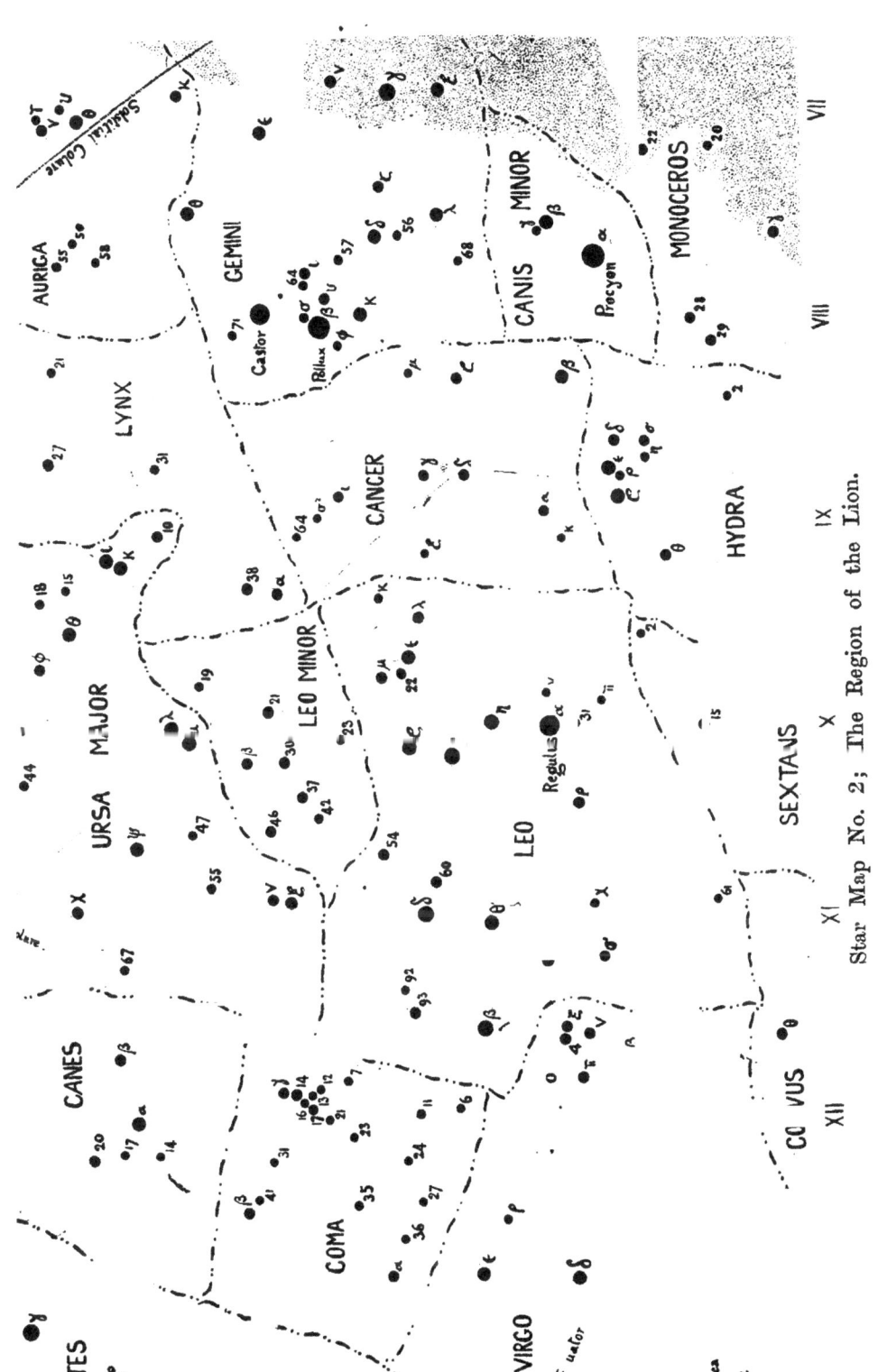

Star Map No. 2; The Region of the Lion.

yellow, whilst the latter is white. Gamma, Zeta and Epsilon are all interesting as opera-glass objects from the companions which a slight optical assistance brings into view.

Leaving the Sickle, we come to the Rectangle, the four stars marking which are of very different magnitudes. Delta and Theta mark the western side; 93 and Beta the eastern; of these Beta is much the brightest, Delta following next. Beta is Denebola,— one of the many Denebs, that is "Tail," which we find in the sky,—and from its companion stars, forms an interesting opera-glass field.

CANCER.

Preceding the Sickle of Leo is Cancer, the smallest and least conspicuous of all the constellations of the Zodiac. Its most significant feature is found in the centre of the group; a pair of stars between the fourth and fifth magnitude, Gamma and Delta, north and south respectively of a misty looking object. These are the twin "Asses," standing right and left of their "Manger," Praesepe.

> " Like a little mist,
> Far north in Cancer's territory it floats.
> Its confines are two faintly glimmering stars;
> These are two Asses that a Manger parts."

Many a young beginner has fancied that in Praesepe he has discovered a new naked-eye comet, but the least optical aid shows it to be a cluster of small stars, and directly Galileo turned his telescope upon it, he detected its nature, counting some thirty stars within its borders. The Asses and the Manger appear to be ancient names, but there are some slight variations in the figure ascribed to the entire constellation, the Egyptians tracing here a scarabaeus, and some of the mediæval astronomers representing it by a lobster or crayfish.

LEO MINOR.

Between Leo and Ursa Major is a modern constellation, called Leo Minor, framed by Hevelius out of the unformed stars which he found in this region. None of its components exceed the fourth magnitude, and it is chiefly noticeable to the naked-eye astronomer as the home of a meteor radiant of the second rank.

LYNX.

The Lynx, lying between Ursa Major and Cancer, is an even fainter constellation than Leo Minor; its two principal stars Alpha and 38 make a visual pair, very similar to the three that have already been noted as marking the plantigrade feet of Ursa Major, and as Prof. Young has suggested, they might well have been taken to have made up the fourth, though, had this been so, our Bear would have been a " high-stepper " of most un-ursine agility.

CANES VENATICI.

Underneath the three stars which make the handle of the Plough, or tail of the Bear, is a bright star, easily recognised from the comparative bareness of the region in which it is placed, which is known as Cor Caroli, " Charles' Heart," so called because Sir C. Scarborough declared that it shone with peculiar brightness the night before Charles the Second made his entry into London on his restoration. This name, however, attaches only to the single star; the constellation like Lynx and Leo Minor, being one of those which we owe to the ingenuity of Hevelius, who named it Canes Venatici, the "hunting dogs." Cor Caroli is a beautiful double star, the components of which are about 20″ apart.

Almost midway between Cor Caroli and Arcturus, but nearer the latter star, is the cluster No. 3 in Messier's catalogue.

COMA BERENICES.

Below Canes Venatici, and immediately to the east of the rectangle of Leo, is a constellation which, though ancient, is by no means one of the original ones. Though it possesses no bright stars, yet on a clear night the region will attract the attention of the sharp-sighted observer, for delicate points and films of light are crowded in it. Serviss writes of it:—

"You will perceive a curious twinkling as if gossamer spangled with dewdrops were entangled there. One might think the old woman of the nursery rhyme, who went to sweep the cobwebs out of the sky, had skipped this corner, or else that its delicate beauty had preserved it even from her housewifely instincts."

The story of its naming is that Berenice, the Queen and sister of Ptolemy Euergetes, vowed her beautiful hair to Aphrodite, should her consort return safely from an expedition on which he had set out. The consecrated tress was, however, stolen from the temple soon after its dedication, and the consequences might have been very serious had not the royal astronomer of Alexandria, Conon, rises to the occasion, by declaring that Aphrodite had caught the tress up to heaven, in proof whereof he pointed out the constellation to the king and queen. Probably, however, the stars in this region had already a half-recognised position as forming a separate constellation, and the quick wit of the astronomer but confirmed a brevet rank.

VIRGO.

The Lion is followed in the Zodiac by the Virgin, which seems almost to lie below the royal beast, for at this time of the year the ecliptic curves downwards more sharply than at any other period, its descending node lying close to the boundary of Leo and Virgo and just within the latter constellation. Virgo, therefore, is easily found when Leo is known, or the old rhyming direction will plainly point it out :—

"From the Pole Star through Mizar glide
　　With long and rapid flight,
Descend, and see the Virgin's spike
　　Diffuse its vernal light,
And mark what glorious forms are made
　　By the gold harvest ears,
With Deneb west, Arcturus north,
　　A triangle appears :"

Denebola of the Lion's tail forming an equilateral triangle with Arcturus and Spica, the principal star of the Virgin.

The chief stars of the Virgin, six in number, make an irregular capital Y, lying on its side, the stem and lower branch of the Y very nearly marking the ecliptic. That great circle is particularly clearly marked out in this portion of the heavens. Delta in the Twins—the bright star below Pollux, and marking that hero's right hand ; Delta in Cancer, the southern of the two Asses ; Regulus in Leo are all almost exactly on the ecliptic, and Rho and Tau, two fainter stars in Leo, carry on the line to the boundaries of Virgo. Within that constellation the line runs a little south of Beta, Eta, and Gamma, which form the right branch of the Y, and a little north of

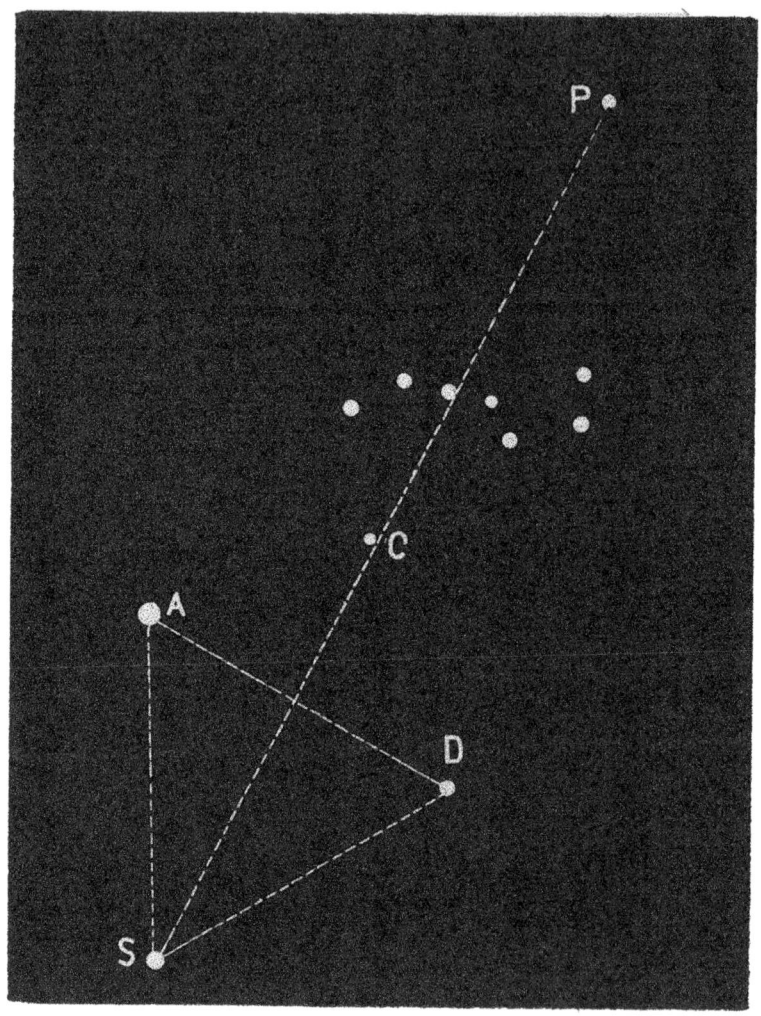

P, Polaris. C, Cor Caroli. A, Arcturus. D, Denebola. S, Spica.

FIG. 15.—Spica and Cor Caroli.

Spica (Alpha Virginis). Whilst a fourth magnitude star, Lambda, as far beyond Spica as Spica is from

Gamma, marks almost the precise point where the ecliptic runs into Libra. The upper branch of the Y is marked by Delta and Epsilon. Gamma, the star which marks where the Y forks, is one of the most celebrated of double stars.

Aratus gives more space to the history of this constellation than to any other. With him, she is Astraea, the spirit of Justice, once in the Golden Age a dweller amongst men. But when an inferior race in the Silver Age succeeded to their fathers, she withdrew to the mountains, and fled thence to the sky when the Brazen Race fashioned murderous weapons and devoured the flesh of plough oxen for their food. The account of her which is still most generally received, is that she represents the wheat harvest; the ear of corn in her hand, which one would have thought a fitter symbol of sowing, being taken as representing the garnered sheaves. But this cannot be the case, for Aratus tells us—

> "As rushing on his prey,
> The lordly Lion greets the God of day,
> When out of Cancer, in his torrid car
> Borne high, he shoots his arrows from afar,
> Scorching the empty fields and thirsty plain,
> Secures the barn the harvest's golden grain;"

proving, as Brown points out, that Spica was not associated originally with the harvest, since this had been already reaped when the sun entered the Lion. A further proof is afforded by the old name of Epsilon Virginis, Vindemiatrix, the "Herald of the Vintage," the vintage necessarily falling considerably later in the year than the harvest.

The constellations of the Zodiac, if intended to mark the several months of the year, should, being twelve in number, stretch each of them over 30° of longitude, neither more nor less. As a matter of fact they are of a most irregular length, Cancer extending only over 18° or 19°, whilst Virgo covers about 50°. At an early period, therefore, the ecliptic was divided into twelve equal portions, not constellations, and having no direct connection with the actual arrangement of the stars, but deriving their names from the constellations which most nearly corresponded to them. These were the Signs of the Zodiac as distinguished from the Constellations of the Zodiac, and the distinction between the two is one that it is important to bear in mind. The months of the year never did, and never could have corresponded with the actual constellations; the Signs, being purely arbitrary divisions, could always be made to correspond with the months. Since, then, the constellation figures are clearly older than the equal signs, it is manifest that none of the many schemes which have been framed to account for the Signs of the Zodiac by the climatic changes of the successive months in this or that country can have any basis in fact. The constellation figures were in existence long before the correlation of signs and months was effected.

The Accadian calendar connects the sixth sign of the Zodiac with Ishtar, "the daughter of heaven," the moon in one aspect, and the planet Venus in her two fold character of morning and evening star in another. Early Christian thought recognized a reference to the promise of "the Seed of the Woman" of Genesis iii. 15, in "the ear of the corn" the Virgin carries in her hand, and in Shakespeare's play of Titus Andronicus, the expression "the good boy in Virgo's lap," refers to the mediæval representation of the sign as the Madonna and Child.

The region of the sky enclosed between the two arms of the Y, and Denebola and Leo, lies near the pole of

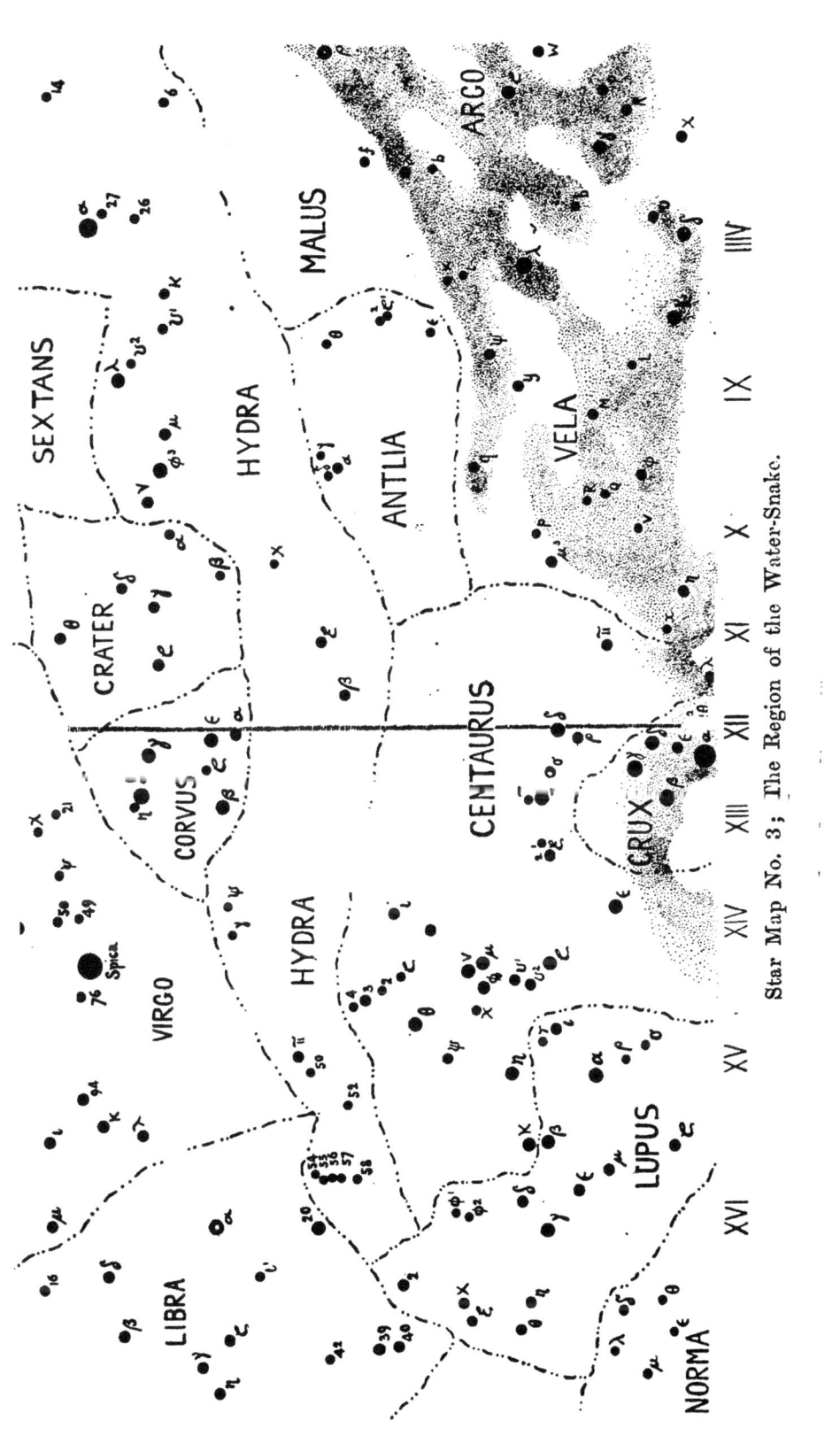

Star Map No. 3; The Region of the Water-Snake.

the Galaxy, and is the wonderful Nebulous Region.   Here these strange bodies are to be found by the hundred, clustering more thickly than in any other portion of the sky.

HYDRA, CORVUS, AND CRATER.

Close below Virgo are two small but fairly bright constellations, the Cup and the Crow; the Cup lying underneath the Virgin's shoulder, the Crow beneath her hand. The latter constellation is very easily found; Delta Corvi forms with Alpha and Gamma Virginis almost an equilateral triangle, and the line from Alpha Virginis to Delta Corvi leads to Gamma Corvi. When on the meridian, two other stars of about equal brightness, Epsilon and Beta, lie below Gamma and Delta and make up with them a neat little trapezium. Beta Virginis and Delta Corvi make a rough equilateral triangle with Delta Crateris, the brightest star in the centre of the Cup, a somewhat fainter group than that of the Crow. Four stars in a semi-circle, of which Delta is the middle, mark the bowl of the Cup. These two little groups are commonly represented as actually intermingled with a huge winding snake, Hydra, the longest constellation in the sky, stretching across some 100° of longitude. Its head begins close to Procyon, under Cancer, and it extends below the zodiacal constellations of Cancer, Leo and Virgo and the greater part of Libra. It has few bright stars, and these not grouped in easily remembered figures; and the great reaches of barren sky it includes seem referred to in the name given to its brightest star, Alphard, The "Solitary." Alphard may be readily found by prolonging a straight line from Gamma Leonis through Regulus, and dropping a perpendicular on it

from Procyon. The myths connected with the three constellations have no very great interest. Brown finds
Hydra, a " storm-and-ocean-monster." "The quick flowing rivers seem to have been compared by the Akkadii
with the swift gliding of a huge glistening serpent, and
so we arrive at the idea of the River of the Snake, which
develops into an Okeanos' stream, like the Norse great
serpent," the Midgard Snake. The Cup becomes thus
a "symbol of the vault of heaven wherein at times
storm, wind, clouds, rain are chaotically mixed"; and
the Crow, or rather Raven, is the constellation of the
Storm-Bird. Carl G. Schwartz, who, at the beginning
of last century interpreted the constellations as being a
sort of symbolical geography of the countries on the
west shore of the Caspian, thought these three constellations represented the petroleum wells of Baku. The
great extent of the Hydra, with its folds and knots,
show, beyond mistake, in his opinion, the slow oily flow
of crude petroleum; the Cup is placed there to indicate
the liquid which would have to be held in a cup or some
such reservoir, whilst the Crow indicates its inky blackness!

Of these three constellations, Crater is perhaps the best
for opera-glass examination, yielding some pretty fields.
Zeta Corvi, the faint star nearly midway between Epsilon
and Beta, shows with the opera-glass as an interesting
little double, whilst a much closer pair will be found
near Beta and slightly preceding it.

BOÖTES.

Arcturus is one of the easiest stars to recognise in the
entire sky. If we start from the Pole Star, we find that
the last star in the Plough Handle leads straight to

Arcturus; or the curve of the three stars of the handle of
the Plough, if continued, seems to bring us round to the
same place; or, as mentioned under Virgo, Denebola
of Leo and Spica of Virgo and Arcturus mark out a
triangle, almost equilateral.

The star owes its name to its nearness to the Bear.
It is Arcturus, the "Watcher of the Bear." It is now the
brightest star in the constellation Boötes, the Herdsman,
but in the catalogue of Ptolemy it is not included in the
actual figure, but is an "unformed" star below him.
There seems to have been some reason for this exclusion,
for Theon and Hesychius call Boötes, Orion, and when
Arcturus is excluded, the principal remaining stars of the
constellation make up a representation, pale and distorted
it is true, but a representation for all that, of the most
glorious constellation of the sky. This circumstance may
explain an allusion in Isaiah xiii. 10, which has puzzled
many commentators, "The stars of heaven and the
constellations thereof." The word "constellations" is in
the plural, and is the same word which is in the singular
in Job ix. 9, Job xxxviii. 31, and Amos v. 8; and
is in each case translated by "Orion" with great
probability. Here then it may stand for the two Orions,
Boötes being one. However this may be, the resemblance
between the two constellations will be near enough to help
the student to trace out the figure. Arcturus stands
nearly midway between the Herdsman's two legs, marked
respectively by the stars Eta and Zeta; above Arcturus
are the three belt stars, Rho, Sigma, and Epsilon, Epsilon
being much the brightest. Above we find Gamma and
Delta marking the shoulders, whilst Beta takes the place
of the cluster of small stars which denotes the head of
Orion.

Delta shows as a double star in the field-glass, a bluish

seventh magnitude companion shining at a distance of a little under two minutes of arc.

CORONA BOREALIS.

Arcturus and Gamma form two points of an equilateral triangle with Epsilon in the centre; the third point is formed by Alphecca, the Broken Platter. The reason of the name is readily seen, since right and left of Alphecca

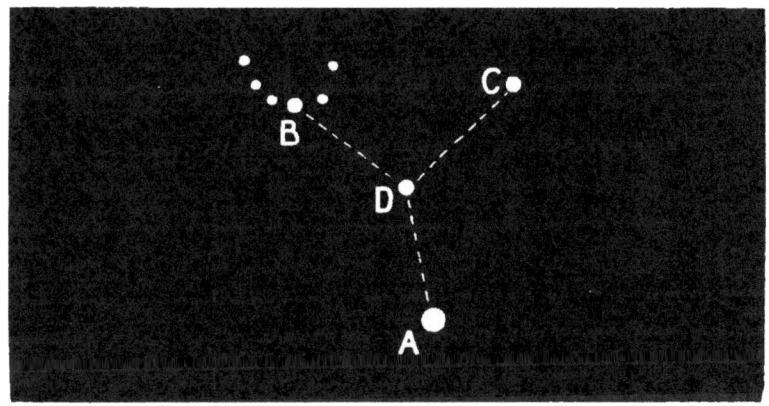

A, Arcturus.   B, Alphecca.   C, Gamma Bootis.   D, Epsilon Bootis

Fig. 16.—Arcturus and the Crown.

are four other stars, two on each side, making up a semi-circle, and suggesting to the old Arabian star-gazers a broken plate held out by a beggar to receive alms. This very sordid title contrasts poorly with its classical name,

> " That Crown which Dionysos placed
> Of Ariadne dead, a glorious sign."

The constellation though so small is, from its shape and its nearness to Arcturus, very easy to find. Or the old rhyme may guide us if we turn back to Virgo, and pick out Epsilon, the " Herald of the Vintage."

" From Epsilon in Virgo's side, Arcturus seek and stem,
   And just as far again you'll spy Corona's beauteous gem;
   There no mistake can well befall e'en him who little knows,
   For bright and circular, the Crown conspicuously glows."

The small size and neat arrangement of Corona make it a pretty object for the opera-glass; and in 1866 it afforded a grand chance for the naked-eye observer. For on the night of May 12th in that year, the constellation suddenly presented an unwonted shape. Epsilon, the star of the five furthest to the east, was overshadowed by a new and bright companion which outshone Alphecca. This was T Coronae, the first "new star" to appear since the invention of the spectroscope. Less brilliant than the new star in Perseus which so suddenly blazed out upon us in 1901, it created, as the first example of the kind that had occurred in the new era of astronomy, an even greater sensation; and the discovery in its spectrum of the bright lines of hydrogen aroused the utmost interest. Four novæ have appeared since the date of T Coronae, including the one so recently discovered by Dr. Anderson. So far, however, as the relatively incomplete observations of its spectrum changes go, they seem to point to T Coronae being a nova of a different order from those which have succeeded it.

SEXTANS.

Immediately under the Sickle of Leo, and enclosed between that constellation, and Crater, and Hydra, is a dull region with a few faint stars, of which only three are brighter than the fifth magnitude. These were combined by Hevelius into the constellation of the Sextant.

CHAPTER V.

THE STARS OF SUMMER.

VEGA, now the zenith star for our latitudes during the midnight of early summer, but which crossed the meridian at midnight in spring at the time when the constellations were mapped out, stands between two groups of stellar figures of most striking significance in the light which they throw upon the origin of the constellations. For to the west of Vega lie five constellations, the Dragon, Hercules, the Serpent-holder, the Serpent, and the Scorpion, which present in double form the Great Conflict; and to the east we have the region of the Birds—three in number—the Swan, the Eagle, and the constellation of which Vega is itself the chief gem, the Lyre, which is represented as an eagle which carries a lyre round its neck.

The old constellation makers have left evident proof in this portion of the sky that they were not working haphazard in the designs they selected for the star groups, and the places which they assigned to them. At midnight at the spring equinox, the Scorpion was for them on the meridian in the south, and the Dragon was in like manner on the meridian low down in the north. Just as they had planted the Kneeler, whom we now call Hercules, upon the Dragon in the north, so they provided another hero, the Serpent-holder, to trample down the Scorpion in the south, and the heads of the two heroes were made up by the stars in the zenith. Both the unknown warriors, therefore, were pictured in those primitive ideas as erect, but for many generations Hercules has been to us hanging head downwards in the sky in the most uncomfortable of attitudes, for our zenith, nowadays, passes nearly through his feet.

It is impossible to suppose that it is due to a mere accidental proximity that these five constellations give us this two-fold picture of a struggle between a man and various malevolent forms. The attitude of the Serpent-holder is the more significant in that one of his feet crushes down the Scorpion's head, whilst the reverted sting of the creature is curled up to wound his other heel, an attitude which recalls most strikingly the prophecy of Genesis iii. 15.

HERCULES.

A line from Gamma Bootis through Theta Coronae, the most westerly of the five stars of the " Broken Platter," brings us, at an equal distance beyond, to Beta in the constellation of the Kneeler. This is not a brilliant constellation, having no stars so bright as the second magnitude, but it can be pretty easily traced out. Taking Beta and the somewhat fainter star Gamma, just below it, as the root, the stars map out the calix of a gigantic lily; Gamma, Beta, Zeta, Eta, Sigma and Tau, six stars in a beautiful curve sweeping round the little constellation of the Crown, forming the western outline of the flower. Hercules is the name now universally ascribed to this

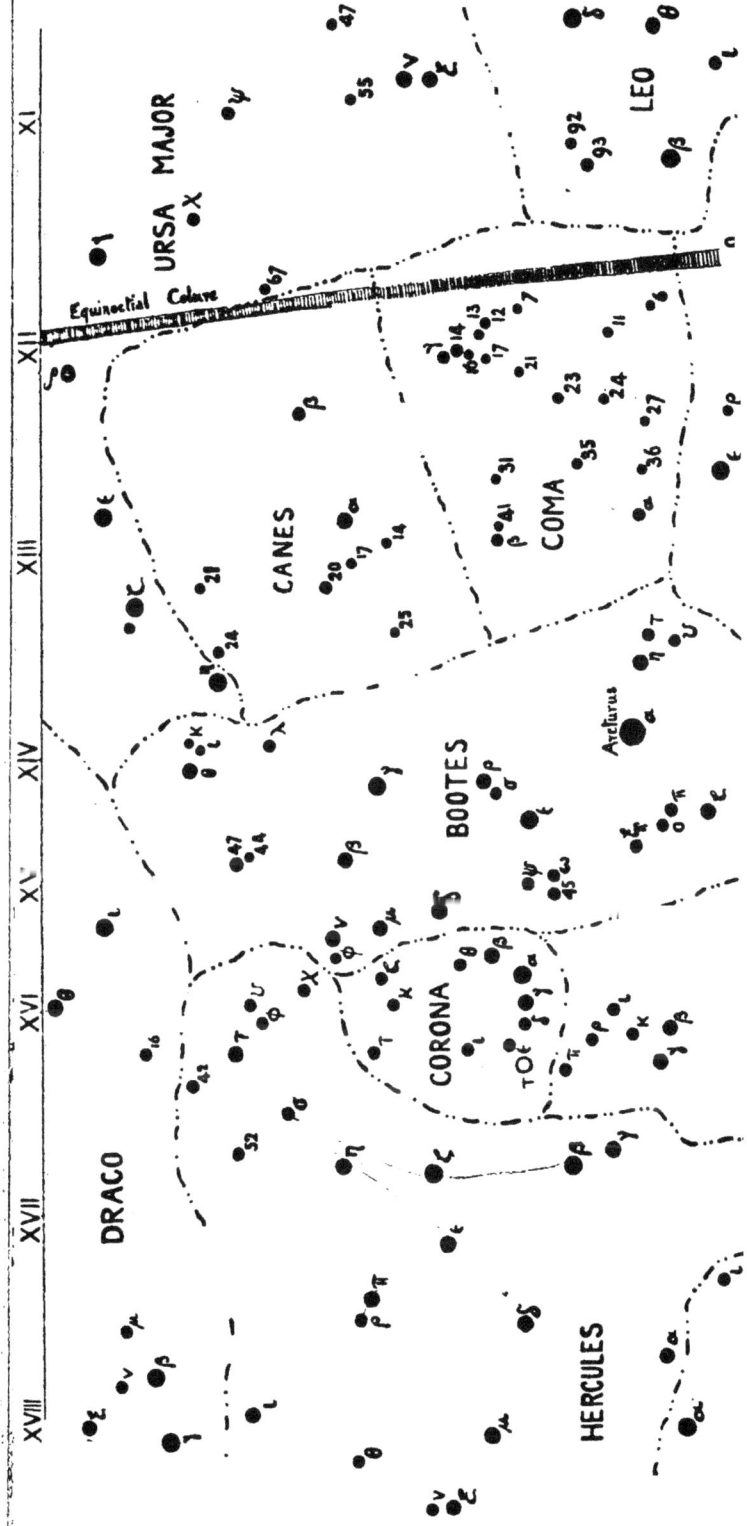

... in the north, so
... shoulder, to tram-
..., and the heads of
... stars in the zenith.

... were pictured
... for many genera-
... head downwards
... attitudes, for
... his feet.

... due to a mere
... stellations give us
... between a man and
... of the Serpent.
... one of his feet
... whilst the reverted
... and his other heel,
... the prophecy of

... Beta Coronae, the
... "Broken Platter,"
... to Beta in the
... not a brilliant
... as the second
... out. Taking
... just below it,
... of a gigantic lily;
... Tau, six stars in a
... constellation of
... of the flower.
... ascribed to this

constellation, but the name was foisted upon it in comparatively recent times.   Aratus sings:—

> "A labouring man next rises to our sight,
> But what his task or who this honoured wight
> No poet tells.   Upon his knee he bends,
> And hence his name, Engonasin, descends.
> He lifts his suppliant arms and dares to rest
> His right foot on the scaly dragon's crest."

The first suggestion that this Kneeler was the great
national Hellenic deity, seems to have been due to Panyasis,
the uncle of the great historian, Herodotus.   In a poem
on the subject of the great national hero, in order to do
him the greater honour he sought to identify him with
the unnamed wrestler of the constellation.   The fact that
despite this effort the identification had entirely failed of
adoption 200 years later, is as near positive proof as we
can get, not merely that it was not known whom the
constellation represented, but that it was known that it
did not represent Hercules.

On the first curve of the lily of Hercules, two-thirds of
the distance from Zeta to Eta, is the great Cluster,
Messier 13, generally considered the finest of the globular
clusters.   It may be picked up in the opera-glass, but its
full glory can only be appreciated by those who have the
command of a first-rate telescope.

The second curve of Hercules runs through the stars
Gamma, Beta, Epsilon, Pi and Iota; Iota making a
diamond with the three stars in the Head of the Dragon,
Beta, Gamma and Xi.   This diamond, Proctor, in his
ingenious but usually quite unauthorised alterations of
the current outlines of the constellations, regarded as
marking the true Dragon's Head.

The third curve of the great lily of Hercules extends
from Gamma and Beta, through a well-marked line

of stars, Delta, Lambda, Mu, and Nu, to the little constellation of the Lyre, the principal star of which is the great blue brilliant Vega, the worthy rival of Arcturus and Capella, if not superior to either.

LYRA.

"There is the Shell but small. And this, whilst yet
  Encradled, Hermes pierced and called it Lyre;
  Fronting the Unknown Form" (*i.e.*, the Kneeler) "he set it down
  When brought to Heaven."

The principal stars of the constellation are very easy to recognise. Vega forms one of the points of a little equilateral triangle, the other two angles of which are occupied by Epsilon and Zeta. Epsilon is to very keen sight a naked-eye double; the opera-glass separates the two stars at once, and no great telescopic power is required to show each star as itself a neat little pair. Zeta marks also the upper angle of a little rhomboid, of which Beta, Gamma, and Delta mark the other angles. Each of these stars is an easy double for the opera-glass; Nu and Lambda being companions to Beta and Gamma respectively. Beta is one of the most interesting of short period variables; its period being two hours short of thirteen days, in which time it passes through two maxima and two minima, the minima being, however, of unequal brightness; but as even when faintest it is of magnitude $4\frac{1}{2}$, it is always well within the grasp of the naked-eye observer.

The Milky Way flows across the S.E. angle of the constellation, and this, with its dazzling leader, its numerous pairs, its beautiful fields and wonderful variable, renders it a fine region for the opera-glass observer. To the naked-eye astronomer, it is also noteworthy as the home of the swift meteors of April 20th—the Lyrids—their radiant point being just on the boundary line between Hercules and Lyra.

The constellation is always shown now as an eagle with
a harp slung round its neck, and the name of the principal
star, Vega, refers to this design, since it comes from the
last word of the Arabic expression, *Al nesr al waki*, the

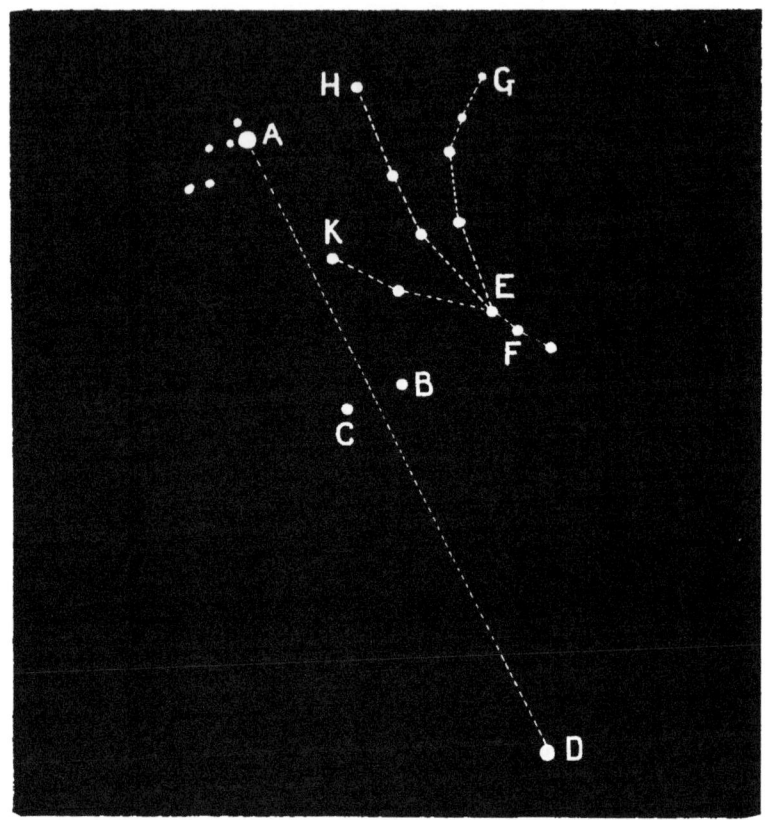

A. Vega. B. Ras al Gethi. C, Ras al Hague. D, Antares. E, Beta Herculis.
F, Gamma Herculis. G, Tau Herculis. H, Iota Herculis. K, Mu Herculis.

Fig. 17.—Lyra, Hercules and Antares.

"falling" or "swooping eagle"; in contrast to Aquila,
the principal star of which we now call Altair, that is to
say, *Al nesr al tair*, the "flying," that is, the "soaring
eagle."

The head of Hercules is marked by a beautiful orange
coloured star, Alpha Herculis, Ras al Gethi, the "head of

the Kneeler," forming the southernmost point of a lozenge of which Beta, Zeta and Delta Herculis are the other three points. Alpha Herculis is notable in the spectroscope as presenting one of the finest examples of the third or banded type of spectrum.

OPHIUCHUS AND SERPENS.

As mentioned above, the old astronomers who depicted the Kneeler in the northern sky kneeling on the Dragon and crushing his head, placed in the southern sky the Serpent-holder, whom we usually call by his Greek name, Ophiuchus, trampling on the Scorpion and strangling the Serpent. The heads of the two heroes, therefore, met together near the zenith, and the chief star of each constellation being in the head of the figure, the two are close together. Alpha Ophiuchi, therefore, is very near to Alpha Herculis; it is two degrees further south, and follows it across the meridian twenty minutes later.

Although the conqueror over the Scorpion, Ophiuchus is not for a moment to be compared with his enemy as a constellation. He covers a great extent of sky, his stars are none of them of the first rank, and are not disposed in any easily followed figures. His principal star, Alpha, Ras al Hague, "the head of the serpent charmer," lies midway between Vega and Antares; as the old rhyme has it:—

"Through Ras al Hague, Vega's beams directs the enquiring eye
Where Scorpio's heart, Antares, decks the southern summer sky."

Ophiuchus is engaged on a double labour. Aratus describes him thus:—

"His feet stamp Scorpio down, enormous beast,
Crushing the monster's eye and platted breast.
With outstretched arms he holds the Serpent's coils;
His limbs it folds within its scaly toils;

With his right hand, its writhing tail, he grasps ;
Its swelling neck, his left securely clasps.
The reptile rears its crested head on high
Reaching the seven-starred Crown in northern sky."

The head of the Serpent is marked out by five stars in
the shape of a capital X, immediately below the semi-
circle of the Northern Crown. The five stars are Beta
and Gamma at the feet of the X, Kappa in the centre,
and Iota and Rho at the top. The small stars clustering near this X of the Serpent's head make an interesting
field to the opera-glass, but the tip of the tail, marked
by the star Theta, is more interesting still. Theta may
be found by drawing a line from Beta Herculis through
Alpha Ophiuchi, and it is situated in a striking channel
in the Milky Way, one side branch of which comes to
an end just on the borders of Ophiuchus.

SCORPIO AND LIBRA.

The Scorpion on which Ophiuchus tramples is a notable
beast, and the claim has been made for him that he
occupies a double portion of the Zodiac. On our modern
maps he lies almost entirely below the ecliptic, which
actually lies for the most part in Ophiuchus, not considered
a zodiacal constellation. This is a striking anomaly if
the limits of the constellation are taken as they are at
present, but if the Scorpion extended over what is now
the constellation of the Balance as well, it would have a
much stronger claim to zodiacal rank than the Serpent-
holder could make. In this case, however, there would be
only eleven figures in the Zodiac, and the enquiry arises—
How many constellations did the original Chaldean
Zodiac contain,—eleven or twelve? The question is a
very important one as bearing on the origin of the

Zodiac, since the twelvefold division is significant of the ancients having determined the length of the year at least approximately before the constellations were mapped out.

The assertion that the Chaldean Zodiac consisted originally of only eleven constellations is made explicitly by Servius the grammarian, in his commentary on the works of Virgil. The latter in his address to the Emperor Augustus, in the first Georgic, suggests that the space which lies between the Virgin and the following Claws, lies vacant for him; "the glowing Scorpion drawing back its arms and leaving for him a more than ample space of sky." Our Greek authorities, however, made the signs of the Zodiac twelve in number, but gave to one of the figures a double space; the Scorpion occupying one sign with its body and another with its outstretched Claws. Brown's explanation of this double honour given to the poisonous creature is very ingenious. Quoting from Aratus, he points out the old legend of how Orion was slain by a gigantic scorpion in punishment for his attack upon Artemis :—

> " And great Orion, too, his (the Scorpion) advent fears.
> Content thee, Artemis ! A tale of old
> Tells how the strong Orion seized thy robe
> When he in Chios, with his sturdy mace
> A hunter, smote the beast to gain Œnopion's thanks.
> But she forthwith another monster bade—
> The Scorpion, having cleft the island's hills
> In midst on either side : This, huger still,
> His greatness smote and slew, since Artemis he chased.
> And so 'tis said that, when the Scorpion comes,
> Orion flies to utmost end of earth."
> (Brown's *Aratus.*)

Orion, the most gorgeous of all the constellations, is, according to this theory, a stellar representation of the

sun; the Scorpion, on the other hand, represents the
night and the power of darkness. "As the huge size
of Orion, *i.e.*, that of the sun as compared with the stars,
is always insisted on, so the scorpions of darkness are of

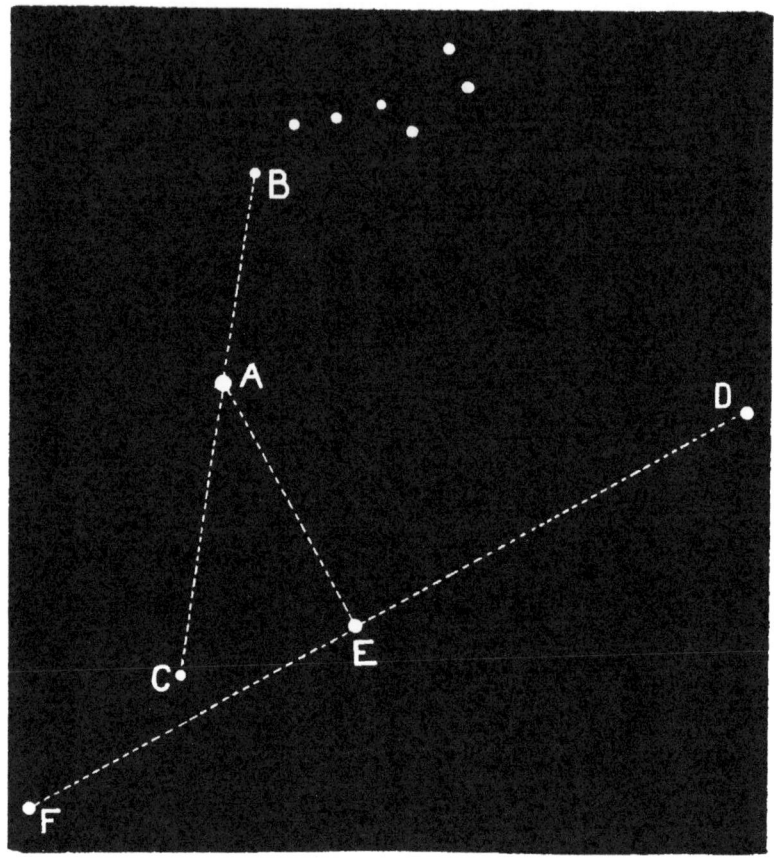

A, Arcturus.   B, Alkaid.   C, Alpha Libræ.   D, Regulus.   E, Spica.
F, Antares.

FIG. 18.—Antares and α Libræ.

colossal size, infinitely greater than the Orion sun." The
gigantic size of the Scorpion, therefore, was insisted on
by bestowing on it a double portion of the Zodiac.

Our present name for the seventh constellation of the
Zodiac, Libra the Balance, we owe to the Romans, Virgil
writing in the first Georgic:—

"Libra die somnique pares ubi fecerit horas."

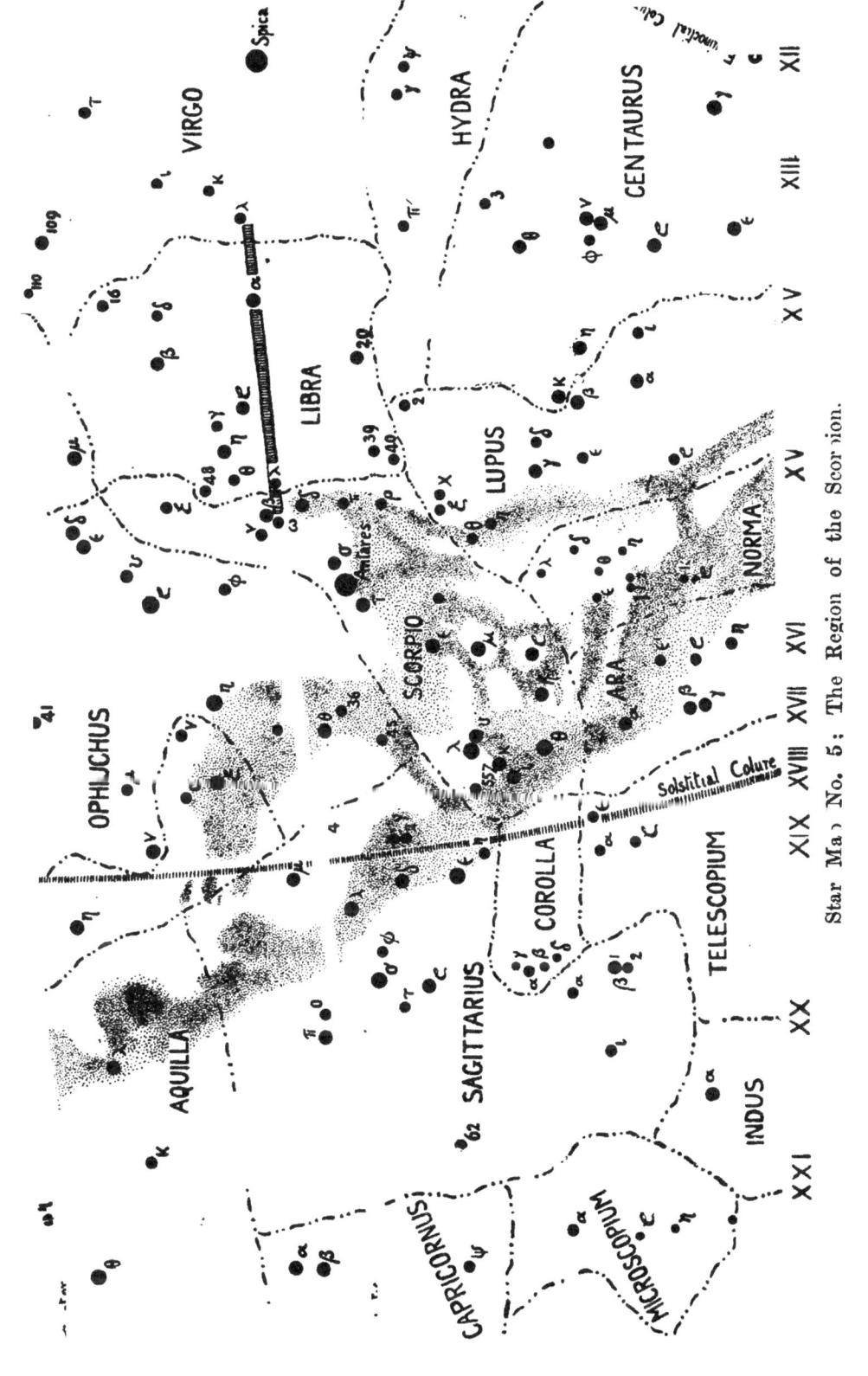

Star Map No. 5: The Region of the Scorpion.

but we cannot now say how far back into antiquity the symbol goes.   Though as Aratus truly says,

"Few are its stars for splendour and renown,"

the constellation Libra can be readily found by an alignment from Arcturus :—

"Where yon gaunt Bear disports a tail, seek Alkaid at its tip,
From thence a ray athwart the space to south-south-east must dip ;
And when Arcturus has been passed prolong th' imagin'd line,
'Twill mark a star, as far again, the first in Libra's sign."

Alpha Libræ is an unlucky star as to its name; it really should be Zuben el Genubi, the southern claw; but it is often called Zuben el Chamali, the northern claw, a title which clearly belongs to Beta Libræ. The star marks the ecliptic almost exactly, and it forms a very pretty double to the opera-glass.

Beta Libræ, forming an equilateral triangle with Alpha Libræ and Mu Virginis, is a star with a markedly greenish colour and interesting history. For Eratosthenes calls it the brightest of all the stars in the Scorpion, that is in the double constellation, and Claudius Ptolemy gives it as equal with Antares. As it is now a full magnitude fainter than Antares it must have faded greatly.

But passing on to the Scorpion proper, we come to the finest region of the sky as seen in southern realms. There is no difficulty in tracing out the Scorpion. Antares, its brightest gem, is almost exactly as far beyond Spica as Spica is beyond Regulus; and Arcturus, Spica and Antares, make up a magnificent right-angled triangle, Spica marking the right angle. Antares stands in a long and beautifully winding curve, made of bright stars, which mark out the body and reverted sting of the animal; and fainter stars to right and left show its legs. "There are few constellations," as Serviss truly remarks, "which bear so close a resemblance to the objects they are named after, as Scorpio. It does not require a very violent exercise of the imagination to see in this long winding trail of stars, a gigantic scorpion, with its head to the west, and flourishing its upraised sting, that glitters with a pair of twin stars, as if ready to strike."

The pair of stars in question are Lambda and Upsilon, Lambda being the brighter. To the north and east of this pair of stars, and about 6° distant from Lambda, are two star clusters about 4° apart—6 and 7 Messier, both well worth examination with the opera-glass. Turning back to Antares, the bright star to the west of it is Sigma, and almost between Sigma and Antares, but a little below, is another star cluster, number 4 in Messier's catalogue. The entire constellation is full of interesting and beautiful fields even for so slight an optical assistance as the opera-glass gives. The two principal stars in the forehead of the Scorpion are Beta the more northern, Delta the more southern, distant from Antares roughly one-third the way to Beta and Alpha Libræ Immediately below Beta is the bright and pretty pair, situated on the ecliptic, Omega Scorpii, whilst Nu Scorpii, a little above and following Beta, is also a double, but requires a more powerful lens to show it as such. Immediately above Antares is 22 Scorpii, a star with two companions; Rho Ophiuchi a little further north, has also a couple, both within the grasp of the opera-glass.

Close to Rho Ophiuchi is the comet-like cluster, Messier 80, called by the elder Herschel the richest and most condensed mass of stars in the firmament.

Following the curve of stars downwards from Antares, —which by the way owes its name to its pronounced red colour, the reddest bright star in the sky, and, therefore, fitly called Antares, the rival of Ares or Mars, —we come in turn to Tau, then after a gap to Epsilon, and then to Mu Scorpii, a lovely pair in the opera-glass, whilst the next star lower down in the curve, Zeta, gives a region of most exceptional beauty to a good binocular. To the English observer, Scorpio hugs the horizon too closely for the full magnificence of the region to reveal itself, but for those who are favoured with a more southern residence, the Milky Way attains here its greatest glory and its most striking complexity of form.

SAGITTARIUS.

Following the Scorpion on the ecliptic comes the Archer, Sagittarius, who carries on the record of the Great Conflict under a new form, for he, like Ophiuchus, is the foe of the Scorpion, and he lays the arrow to the string to slay it. Sagittarius, when on the meridian, is almost entirely above our English horizon, but it lies so low that it is perhaps less familiar to us than any other of the zodiacal signs, for though Scorpio does indeed lie lower still, its brilliance has made it better known. Still, there is no difficulty in recognising it on a very clear night, at its culmination, which takes place at midnight at the end of June. The old rhymester directs us—

"From Deneb, in the stately Swan, describe a line south-west
    Through bright Altair in Aquila, 'twill strike the Archer's breast."

Or, more strictly speaking, his shoulder, marked by the bright star, Sigma. A little in advance of Sigma are five bright stars in an undulating line on the eastern branch of the Milky Way, which here suffers one of its numerous divisions. Proceeding from the most northerly of these downwards, they are lettered Mu, Lambda, Delta, Epsilon, Eta, and mark the position of the Archer's Bow. A pair of stars, both bearing the

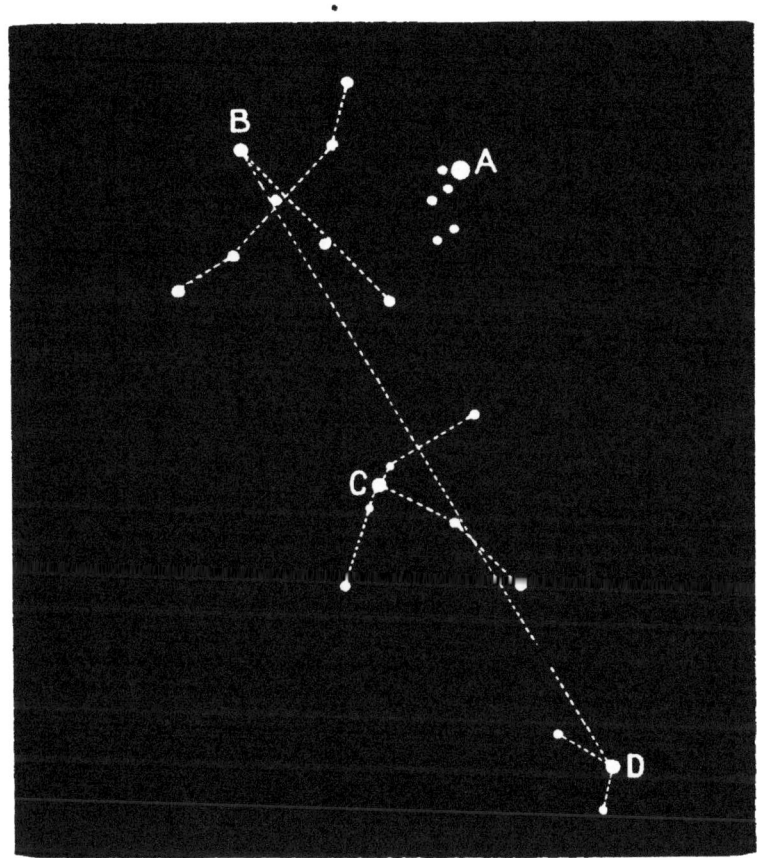

A, Vega. B, Deneb Adige. C, Altair. D, Sigma Sagittarii.

FIG. 19.—Cygnus, Aquila, and Sagittarius.

letter Gamma, a little in advance of Delta, marks the point of the Arrow which the Archer is discharging at the Scorpion, whilst Zeta, a bright star a little below Sigma, marks the wing of the Arrow. A little triangle of stars, Xi, Omicron, Pi, mark the neck of the figure,

and practically exhaust the list of its brighter stars.
Alpha and Beta, the latter a wide double star to the
eye, are in one of the hind legs of the Man-horse, but
are below our English horizon.

But though Sagittarius is not a distinguished constellation, viewed as a whole, it is very rich in objects of
the greatest interest in opera-glass or telescope.   Mu,
the upper horn of the Bow, is the centre of a region
as rich in star clusters as the nebulous region in Virgo
is in nebulæ.   A wonderful object, number 8 in Messier's
catalogue, forms a rhomboid with Mu, Lambda and
Delta.   North of Mu lies number 24 of the same catalogue, a star cluster quite unlike Messier 8, but almost
as attractive.   Passing upwards in the same straight
line, we come to Messier 18, then Messier 17, the famous
"horse-shoe" nebula, and a little further off, Messier
16.   These clusters are the principal objects in the little
modern constellation, Scutum Sobieski, an asterism
which Hevelius devised to celebrate the valiant John
Sobieski, king of Poland, and deliverer of Europe from
the Turk.   Proctor and some other modern map makers
omit the constellation entirely, and for the sake of simplicity it is well that it should be so.   As designed,
it filled a small triangular space between Serpens, Aquila
and Sagittarius.   It was practically entirely enclosed
within the borders of the Galaxy, and contains but a
single notable star, the variable R. Scuti ; but its wealth
of telescopic stars, clusters and nebulæ is most remarkable.   Sir William Herschel estimated that in five square
degrees of space it contained one-third of a million of
stars.   Of its clusters the most wonderful is just visible
to the naked eye, and is Messier 11, the "Flight of Wild
Ducks," on the north-eastern border of the constellation.

CORONA AUSTRALIS.

The fore-feet of Sagittarius are often not shown in the designs of this constellation, the place where they should come being occupied by the Southern Crown.

" Other few,
Below the Archer, under his fore-feet,
Led round in circle, roll without a name."
(Brown's *Aratus*.)

The constellation lies below our English horizon, with very little to mark it from any point.

In the star maps, Proctor's suggested name for the constellation has been given to it—Corolla, the Wreath — so as to distinguish it the better from Corona, the Crown, to which Bootes stretches his hand.

CYGNUS.

Returning to Vega, the key star of this region of the heavens, we find ourselves in the home of the Birds, for close to the Eagle which carries the Lyre are seen the Swan and the Eagle. It is a feature of the primitive constellations which, whatever its significance, cannot escape the most casual notice, that many of the stellar forms, indeed, most of them, are duplicated, and when thus repeated the similar figures are, as a rule, not widely separated but placed close together.

In this region of the sky Aratus tells us,

" There is in front another Arrow cast
Without a bow ; and by it flies the Bird
Nearer the north.   And nigh a second sails
Lesser in size, but dangerous to come
From ocean when night flies ; the Eagle named."

In the midnights of early June, the great stream of

the Milky Way crosses the sky from due south to due north, not passing, however, through the zenith, but somewhat to the east of it.   Right in the centre of this magnificent arch, forming its very keystone, is the constellation of the Swan, easily found from its neighbourhood to Vega, Alpha Lyrae.   The figure of the " Bird," or as we now know it, the " Swan," may be easily traced out.   A long undulating line of bright stars lies parallel to the axis of the Milky Way, skirting the western edge of the great channel which here divides it.   This represents the outstretched neck, body, and tail of the flying Swan.   Crossing it at right angles is another undulating line of stars which represents the outstretched wings of the flying bird.   The whole constellation has often been termed from its shape the " Northern Cross."   Beta Cygni marks the extreme tip of the Swan's bill, and lies about as far beyond Gamma Lyrae as Gamma is from Vega.   Its name is usually given as Albireo, but the meaning and derivation of the word is obscure, and is almost certainly due to a mistake.   The Arabic name is Al minkar al dajajah, the " Hen's Beak."   It is one of the loveliest double stars in the entire heavens; the principal star, of the third magnitude, being topaz yellow, the companion, of the seventh magnitude, sapphire blue, and the distance, 35", being within the power of a field-glass.

Gamma, the bright star which marks the intersection of the cross, is the centre of a most interesting region.   The whole extent of sky from Beta to Gamma is perhaps the richest in the northern heavens, and Gamma itself is in the midst of rich streams of small stars, interspersed with some strikingly definite dark lanes. Of the transverse beam of the cross, Epsilon marks the eastern arm, Delta the western, and from Gamma to

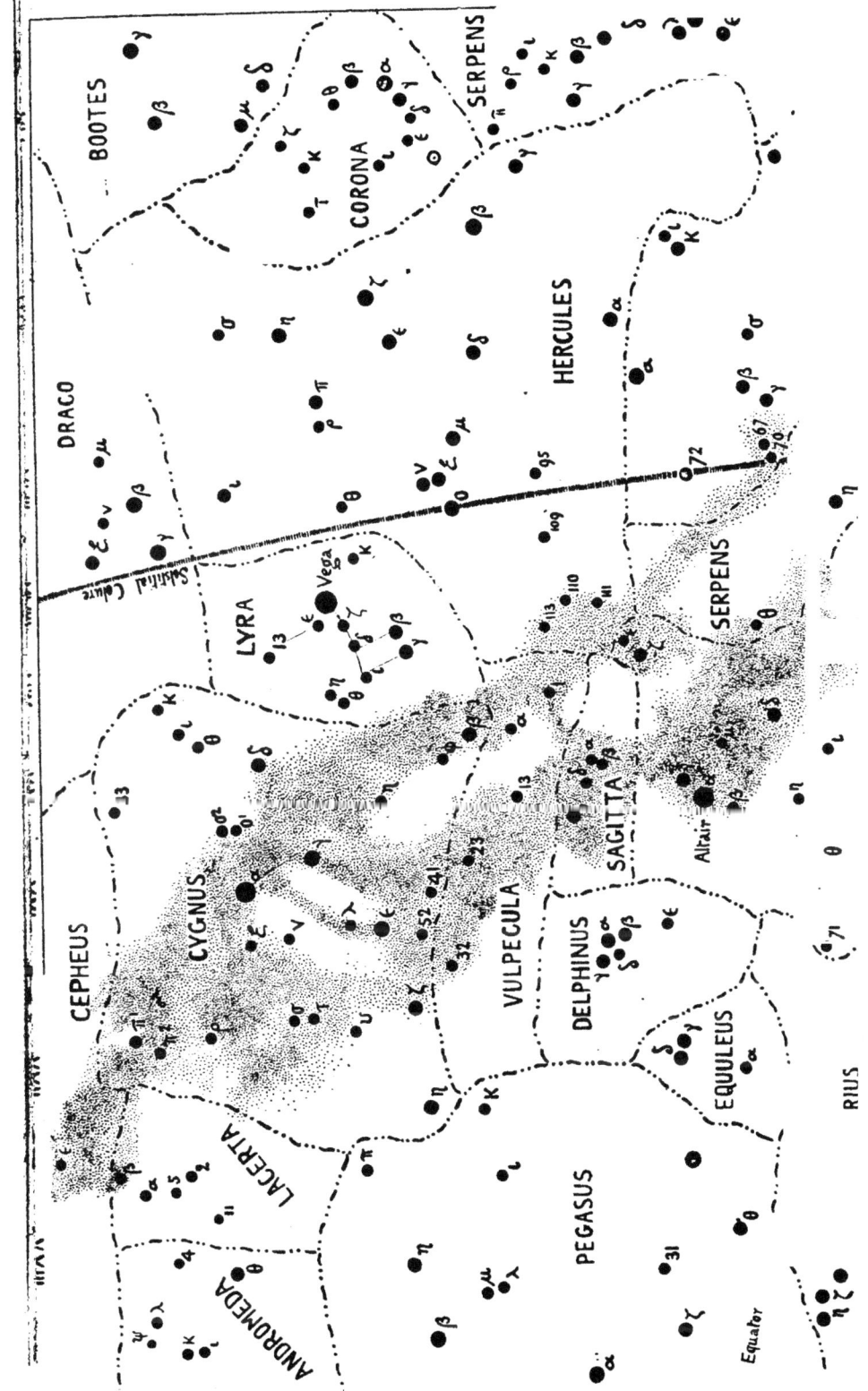

Epsilon we find one of the most remarkable gaps in the Milky Way, the "Coal Sack." Alpha has been called Deneb Adige, the "Hen's Tail," or Aridif, the "hindmost" or "follower," both titles appropriate enough to its place. The entire region of the constellation is full of interest and beauty, whatever the optical power with which it is examined, from the naked eye up to the greatest telescope. One of the many interesting objects in the region is Omicron. To the eye a double star makes a trapezium with Alpha, Gamma and Delta; the brighter of the members of this double, Omicron, in the field-glass will be seen to have two companions, one on each side, both of them blue, whilst the chief star is orange.

AQUILA.

Aratus refers to the constellation Cygnus as the "Bird," without naming its species, but Eratosthenes defines it as the "Swan," and its length of neck well agrees with the identification. Dr. Lamb's somewhat doggerel rendering of Aratus enlarges upon his author in this connection to bring in absolutely without warrant the story of Leda and Zeus. The second bird is, however, plainly identified as the Eagle, and its chief star, Altair, forms the third point of a roughly equilateral triangle, the other two angles of which are marked by Vega and Alpha Ophiuchi. It is also easily identified by the two smaller stars, Gamma and Beta, above and below it; these making with it a characteristic figure of three bright stars in a straight line, on the borders of the Milky Way. The three stars bear the following names:—The middle one, Alpha, is Altair the "Bird," that is the "Flyer"; the "Soaring Eagle," as contrasted with Vega, the "Swooping" or "Falling Eagle"; the southern star is Beta, Alshain, the "White Falcon"; and Gamma, the northern star, Tarazed, the "Robbing one."

The rest of the constellation can be made out without much trouble, but the figure is by no means so good as that of Cygnus. Two stars, Zeta and Epsilon, near together, mark the tip of one wing in the north-westerly direction, and a very much fainter pair, 70 and 71, mark the tip of the other wing, about the same distance on the other side of Altair. Proceeding from Gamma down the Milky Way, we find Mu, Delta and Lambda, reproducing roughly the arrangement of stars which marks so clearly the neck and head of the Swan. Following the line of the three stars, Alpha, Beta, Gamma, we find they point downwards to a bright star, Theta; between this and Delta, but nearer to Theta, is Eta, one of the regular variable stars of short period, visible in all its phases to the naked eye, its period being one of four hours over the week.

SAGITTA.

The quotation from Aratus given a couple of pages above, refers to a little constellation which in a certain sense is the most interesting in the entire sky, the constellation of the Arrow. Possessing only five little stars of the fourth magnitude, and extending in a narrow line, but 4° in length, increased by the moderns to 10°, it is nevertheless one of the oldest constellations, being mentioned three times by Aratus in his celebrated poem, and having its five principal stars duly catalogued by Ptolemy. The history of the Arrow was lost even in the time of Aratus. It was not shot by

Sagittarius the Archer; so much is quite clear, for it is flying high above his head and in the opposite direction to that in which he is shooting. This is the explanation of the phrase—

" Another Arrow cast
Without a bow; "

nor can either of the three heroes, who are near at hand, the Herdsman, the Serpent-holder, or the Kneeler have despatched it. The Herdsman grips his crook, the Serpent-holder has both hands full of the twining snake, and if we accept the guess of Panyasis that the Kneeler was really Hercules, Germanicus tells us that one hand held a club, the other a lion's skin.

Parallel to the Arrow, and of not much larger extent, is the modern constellation, Vulpecula, framed by Hevelius in 1690. Its principal interest to the naked-eye observer is the meteor stream which radiates from it in the latter half of June, and to the telescopic observer the celebrated Dumb-Bell nebula, just visible in the field-glass.

CHAPTER VI.

The Stars of Autumn.

THE evidences of design and of connected thought which the stars of summer supply, in the threefold representation of the Great Conflict and in the group of the Birds, are even more distinct when we come to the stars of autumn, presided over by Cassiopeia. For the southern horizon is occupied at this season of the year by a chain of seven constellations, all marine in character, and Cassiopeia herself is a member of a cluster of constellation figures, five in number, which, unique among the stellar designs, sets forth a distinct and well-recognised story. These are the five constellations which, together with Cetus, preserve to us the legend of Perseus and the maiden whom he delivered. The story, as it has come down to us from Greek sources, is one beloved of romancists in all ages and in all lands. A lovely maiden, innocent herself of any fault, is yet condemned, in order to expiate the offences of her parents, to be exposed to some terrible disaster. Her case seems beyond hope or help, when, at the very crisis of her fate a young hero, who has already abundantly proved his mettle in other fields, appears on the scene. Her beauty and her distress alike appeal to him; and to his victorious powers, her deliverance is a light task. The threatening monster is easily disposed of, and what promised to be a grim and terrible tragedy, ends with triumph and rejoicing to the sound of wedding bells.

PERSEUS AND ANDROMEDA.

It may be, as Brown assures us, that we have in the Andromeda legend but another version of the all-pervading solar myth. Perseus may be Bar-Sav, the son of hair, that is to say, the solar Herakles clad in his lion's skin, and Andromeda, his bride, the rosy red dawn; but if so, the dead myth has passed through minds who could fill it with a human interest, and so imbue it with the spirit of life. As in the story of Pygmalion, it may be that that which was cold and dead was the original; but surely for us, as for him, the living Galatea is not only more worthy, but is more real and true than the lifeless marble whose form she bore. So, we may still look upon Andromeda and Perseus as no mere abstractions of natural phenomena, but as the innocent persecuted maiden and her gallant deliverer; the old romance, ever new and ever true throughout the ages of the world's long history.

Two of the five constellations, Cassiopeia and Cepheus, were described in a preceding chapter. The chief stars of the group now under consideration are easy to find when once we have found Cassiopeia. When Cassiopeia is in the zenith, then may be seen, high up in the sky, but slightly to the west of the meridian, four stars marking out the corners of a figure which is generally known from its shape as the square of Pegasus. Taking the two upper stars of the square, which are now known as Beta Pegasi and Alpha Andromedae, and proceeding eastward, we find three bright stars at about the distance apart from each other that the stars of the square are, which are respectively Beta and Gamma Andromedae and Alpha Persei. The latter is readily found. The Milky Way streams down from the zenith, where Cassiopeia is seated, to the east point of the horizon; and along its axis, a chain of bright stars run down from Cassiopeia. The brightest of these is Mirfak, *i e.*, "elbow," Alpha Persei, marking the elbow of the hero as the word denotes. These three bright stars form the upper points of a greater W just below the smaller, more distinct W of Cassiopeia. The lower points of the W are marked by two stars, Beta Persei, better known as Algol, the "demon" star, so called from its

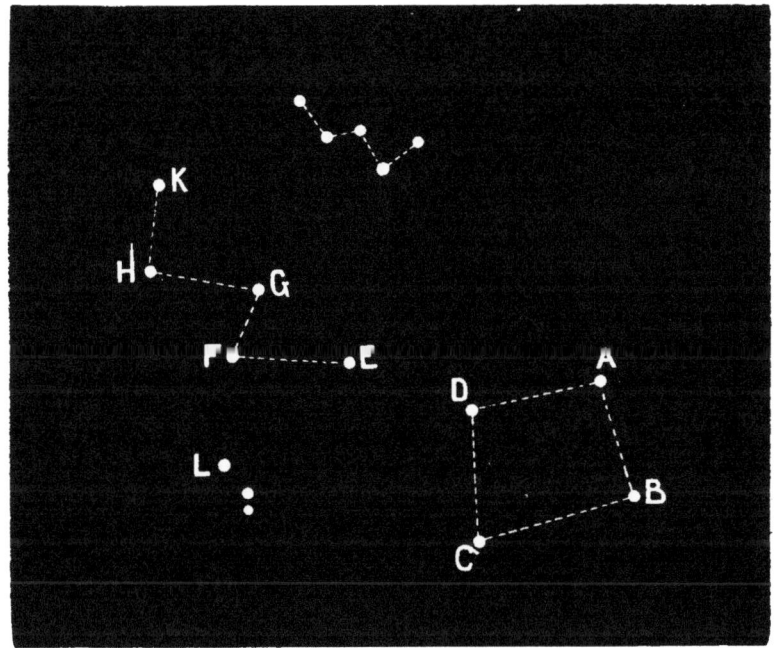

A, Beta Pegasi. B, Alpha Pegasi. C, Gamma Pegasi. D, Alpha Andromedae. E, Beta Andromedae. F, Beta Trianguli. G, Gamma Andromedae. H, Beta Persei. K, Alpha Persei. L, Alpha Arietis.

Fig. 20.—The Greater "W" and the "Square."

wonderful variation, and Beta Trianguli, the brightest star in the small but ancient constellation of the Triangle.

TRIANGULUM.

This little constellation, insignificant as it is in itself, is of great importance, from the evidence which it supplies that the ancient constellations were the result of deliberate design and forethought. Brown's remarks on this point are so just that I may be forgiven for quoting them. Referring to the type of explanation popular in a certain school, he says:—"They would say that someone noticed these stars, saw they resembled a triangle, called them the Triangle, and everyone else followed suit; a pretended explanation which merely repeats the fact that such a constellation exists." It is clear, as Brown further points out, that there are hundreds of stars which might have been combined in triangles, and would have equally suggested such a figure. Not a few would have suggested it much more strongly; as, for example, the stars in the head of Taurus. The selection, therefore of these by no means conspicuous stars to form the constellation of the Triangle is a strong indication that, not only the designs, but their positions were matters of definite purpose to the old constellation makers.

The principal characteristics of the three constellations, Andromeda, Pegasus and Perseus, as they are now shown in our star atlases, are preserved for us in the description of Aratus. Of the first he says:—

"Her garlanded head,—her shoulders bare admire—
Her diamond sandalled feet,—her rich attire;
She still in heaven her captive form retains,
And on her wrists still hang the galling chains."

PEGASUS.

His description of the constellation of Pegasus is one

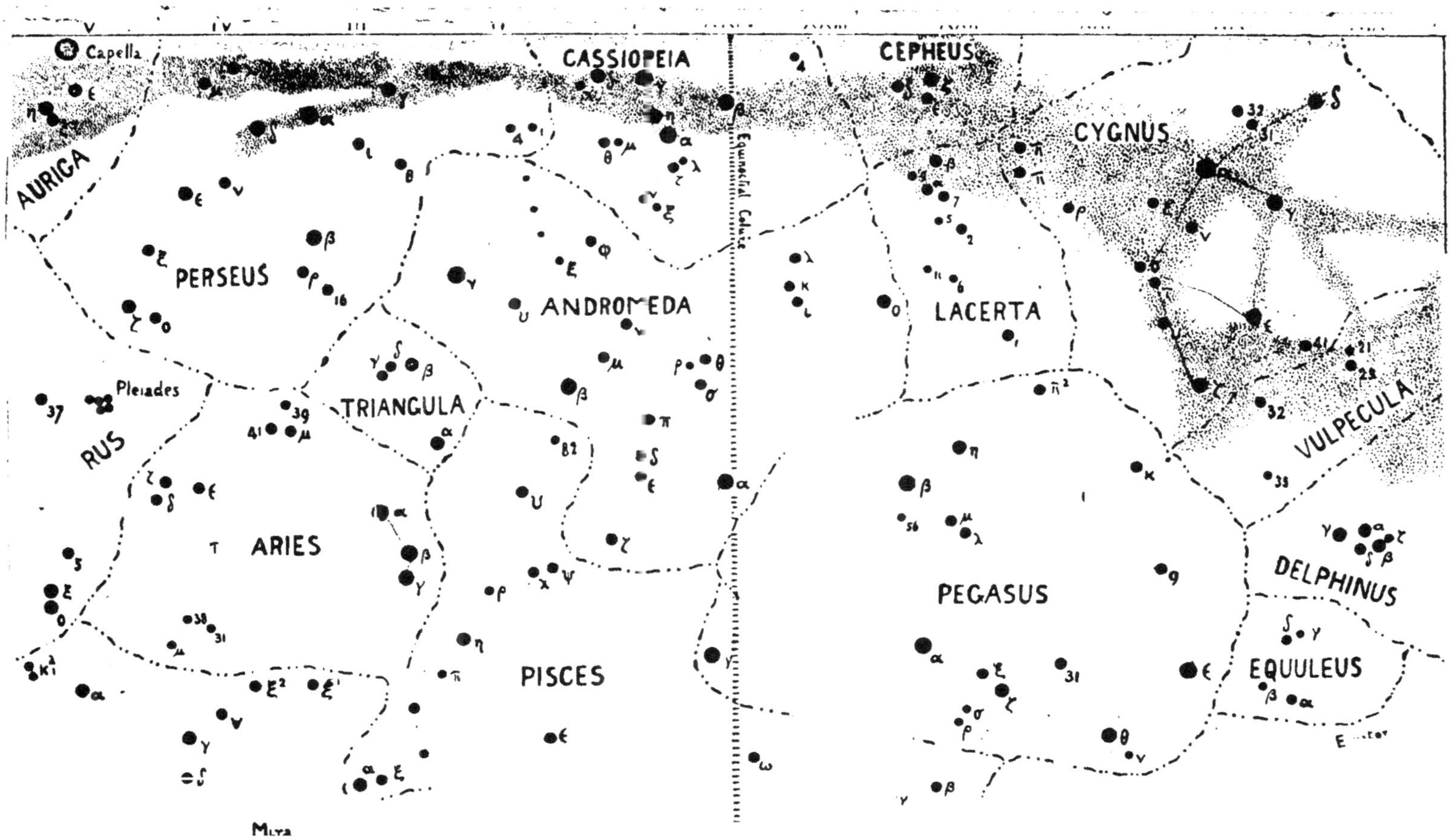
Capella
AURIGA
PERSEUS
Pleiades
RUS
CASSIOPEIA
ANDROMEDA
TRIANGULA
ARIES
PISCES
Equinoctial Colure
CEPHEUS
LACERTA
CYGNUS
VULPECULA
PEGASUS
DELPHINUS
EQUULEUS
Mira

of the fullest and most detailed of any. There is no need, however, to quote the whole, but he calls attention to the circumstance that the star, now known as Alpha Andromedae, the upper left hand star of the square, is common to the two constellations. Its name, Alpheratz, "the Horse," perpetuates the same tradition—

> "Close and above her head the wondrous steed,
> With hoof and wing exerts a double speed.
> So close they meet, one brilliant star they share;
> Its body it adorns, and decks her hair.
> His side and shoulder with three others graced
> As if by art at equal distance placed."

That is to say, these three with the star common to the two figures make up the Square.

Of Perseus, Aratus says, after referring to Andromeda :—

> "Her anxious eyes
> Gleam bright with hope; beneath her Perseus flies,
> Her brave deliverer—mighty son of Jove—
> His giant strides the blue vault climb and move
> A cloud of dust in heaven. His falchion bare
> Reaches his honoured step-dame's golden chair."

The "cloud of dust" alluded to is the Milky Way, on an arm of which, Mirfak, the chief star of the constellation, stands. This "dust" round Mirfak supplies for an opera-glass perhaps the finest field in the entire sky; the whole region being full of winding streams of stars of the most attractive form. Moving upward from Alpha towards Cassiopeia, we pass through Gamma and Eta Persei. From Eta Persei, half-way to Delta Cassiopeia, lies a cloudy object, the great cluster of Perseus, one of the finest sights the heavens have to present. Then, again, in the centre of the triangle formed by, Gamma Persei, Gamma Andromedae and Algol, lies the comet-like cluster, 34 Messier. Crossing over to the constellation of Andromeda, Gamma Andromedae marks the neighbourhood of the shower of meteors associated with Biela's comet, and now encountering our earth about November 23. Passing on to Beta Andromedae, we find it the starting point of a line of three stars pointing upwards towards Cassiopeia. The next of these stars to Beta is Mu, the third Nu, close to which, towards the east, lies the great nebula of Andromeda, 31 Messier—after the great nebula in Orion, the finest example of that order in our skies.

The entire region of Perseus repays examination with the opera-glass, and of Andromeda the region nearest the Milky Way. Pegasus is much less interesting, but possesses a naked-eye double in Pi, the star in the horse's hoof.

When the second pair of the stars which compose the Square of Pegasus is on the meridian, the entire southern horizon from the south-west almost to the east, is held by a succession of designs, all of which represent either fishes or marine animals, or else streams of water.

The first token of these marine symbols occurs in the pretty little constellation of the Dolphin that coils itself behind the outstretched wing of the Eagle. Then follows the fishtail of Capricornus. Next we have Aquarius, with the broad stream flowing from his ewer, and the Southern Fish at his feet. Aquarius is succeeded on the Ecliptic by the long constellation of Pisces, a pair of fishes united by a waving riband. Below this group two other constellations repeat in more terrible form the design of the water-pot of Aquarius. A huge marine dragon known to us to-day as Cetus, the "Whale," but traditionally rather of saurian form, like the

"Monstrous eft that of old was lord and master of earth,"

pours forth from his mouth a vast bifurcating flood, which sweeps down below the horizon.

DELPHINUS.

The first member of this series lies near Pegasus, and is easily found. A line from Alpha Ophiuchi through Zeta Aquilae, and another from Alpha Cygni through Epsilon Cygni, will meet together in a pretty little constellation, which, once picked out, can never be forgotten, its leading stars being so nearly equal in magnitude and so close together. This is the constellation of the Dolphin, containing ten stars in Ptolemy's catalogue. Two of these are a little brighter than the fourth magnitude, and seven others range from that down to the fifth. Alpha, Beta, Gamma and Delta form a compact little lozenge, the straight line of Gamma and Delta being continued on by Theta and Epsilon. Though the Dolphin is one of the ancient constellations, the names attached to the two principal stars are quite modern, and are due to a piece of very clumsy humour on the part of Piazzi, the Sicilian astronomer. In his catalogue he introduced for these two stars the names Rotanev and Svalocin, names which gave a good bit of trouble to etymologists until it was seen that they were simply the name of Piazzi's assistant, Niccolo Cacciatore, latinized and spelt backwards.

CAPRICORNUS.

Capricornus is the next zodiacal constellation to Sagittarius, and is small and easily found. Just as Sagittarius is a centaur or man-horse, so Capricornus is almost invariably a goat-fish. The goat has usually been explained as signifying the sun at the winter solstice, seeing that after that season has passed, the sun begins again to move upward in the sky; the rock-haunting goat or ibex being adopted as the symbol of the climbing motion of the sun, whilst the fishtail pointed to the rains and floods of midwinter. We know, however, that the constellations were mapped out many centuries before the winter solstice fell in Capricorn, and that the explanation, however ingenious, was but a late guess, made when all actual recollection of the meaning of the sign had been lost.

Capricorn may be found by drawing a straight line from Vega through Altair. Omega Capricorni lies just as far below Altair as Vega lies above it, and marks one knee of the kneeling goat. Psi, immediately above, lies almost on the straight line from Altair, and Alpha and Beta, which mark the root of the horns and the eye of the animal, are but little in advance of the line, only considerably higher up. Alpha is Algedi, which simply means "the goat"; Beta is Dabih, "the slaughterer." The first of these stars is a visual double, and it is interesting to note that it has become so comparatively recently. The two stars have no real connection with each other, and their proper motion is carrying them apart by something more than a minute of arc in a thousand years. Beta Capricorni is also a beautiful double in the opera-glass, the fainter star being of a sky-blue tint. The only other stars of any great brightness in the constellation are Zeta, Gamma and Delta, which mark the fishtail. Delta is indeed the brightest star in the whole asterism, and bears the name Deneb Algiedi, the "tail of the goat" Following Zeta by about the same distance that Alpha is from Beta, the opera-glass will show as a faint point of light, 30 Messier, a large cluster of remarkable richness.

H

AQUARIUS.

" Down from bright Vega, cast your glance across the Dolphin's space,
 Then just as far again you'll find the Waterbearer's place."

Aquarius is one of the straggling constellations, and
pretty nearly overwhelms Capricornus.   He has been
figured from time immemorial as a man pouring out a
stream of water from a pitcher ; but for some reason,
which is now lost to us, his right arm is stretched backwards to the fullest extent possible so as to reach over
almost the entire length of Capricorn.   The figure is by
no means clearly marked out in the main.   The stream
from the pitcher can be traced in the number of faint
stars in the eastern portion of the constellation, which
lead downward in wavering curves to Fomalhaut, a star
of the first magnitude and one of the four ancient Royal
Stars.   Dwellers further south can recognise Fomalhaut
without any difficulty, since Achernar lies just mid-way
between it and Canopus.   But at midnight during the
first few days of September it is impossible for
English observers to mistake the star ; it lies low down
on our southern horizon without any serious competitor
near it.

Its name Fomalhaut simply means " the Fish's
Mouth," for strangely enough through all the long
centuries that the starry symbols have come down to
us, Aquarius has always been shown as pouring forth
his stream of water into the mouth of a fish ; surely
the strangest and most bizarre of symbols.

Fomalhaut, Beta Capricorni and Alpha Aquarii form
the points of a triangle which is nearly equilateral.
Alpha Aquarii is known as Sadalmelik, the "fortunate
star of the King"; Beta Aquarii, one-third of the way

from Alpha Aquarii to Beta Capricorni, is Sadal Sud, "the luckiest of the lucky," supposed to refer to the good fortune attending the passing of winter. Alpha and Beta mark the two shoulders of the Waterpourer, and three bright stars near Alpha, Gamma, Eta, and Pi, with a fourth, Zeta, almost in the centre of the triangle, mark the body of the pitcher from which Aquarius is pouring.

The outstretched arm of Aquarius is marked by a slightly curved line of stars extending from Beta Aquarii to Alpha Capricorni. Two fairly bright stars, Mu and Epsilon, near Alpha Capricorni, give the place of the Waterbearer's hand. In the middle of the arm is Nu, a much fainter star, and about a degree and a half preceding it. Just barely within the power of an opera-glass to reveal it as a faint point of light, is one of the most wonderful of the planetary nebulæ. Midway between Alpha and Beta Aquarii, but above the line joining them, is M. 2, in the head of the Waterpourer; "a heap of fine sand" where each grain is a sun.

CETUS.

The figures which tell the story of Perseus and Andromeda are connected with the seven watery constellations by Cetus; the enemy from which Perseus rescued the maiden, and itself emphatically the Sea Monster. Nor is the connection limited to this one design, for the constellation of Pisces has been plainly used to carry on the thought of the hostility between Cetus and Andromeda, which might otherwise have escaped recognition owing to the distance which lies between them in the sky. The Stream, Eridanus, too, which Cetus appears to have poured out of his mouth, is evidently an integral part of the story

as we find it in the heavens, though the old Greek legend
has lost trace of it. The five constellations therefore of
the Royal Family, as they have been called, have a
distinct connection with some at least of the seven
marine groups. But at all events the gathering together
of these seven, all of the same watery character, all in the
same quarter of the heavens, no dissimilar forms being
permitted to intervene and break the chain, is a very
striking circumstance.

It is impossible to suppose that the association of
these seven watery constellations in such close connection with each other can be merely accidental. We
may dismiss at once the idea that they have any special
reference to the rainy season, for an eighth water constellation is supplied us in Hydra, the Water Snake,
which, with the seven just named, very nearly completes
the circuit of the sky. We should therefore have to
conclude that we were dealing with a climate which was
rainy throughout the year, a circumstance not likely to
be symbolized in this particular fashion.

Anything like an adequate discussion of the true
significance of this grouping would lead us too far
astray from our present purpose. Greek mythology saw
in Cetus the monster which Neptune sent to destroy
Andromeda, in punishment for the pride and insolence
of her mother Cassiopeia, and this, of course, is the
account of Aratus.

> "Mark where the savage Cetus, crouching, eyes
> Andromeda, secure in northern skies;
> The Fish and horned Ram his progress bar,
> Nor dares he pass the track of Phœbus' car."

Brown identifies Cetus with the Euphratean Tiamat,
the spirit of Chaos, the enemy of the beneficent gods,
and opposed to law and order; and he notes that the

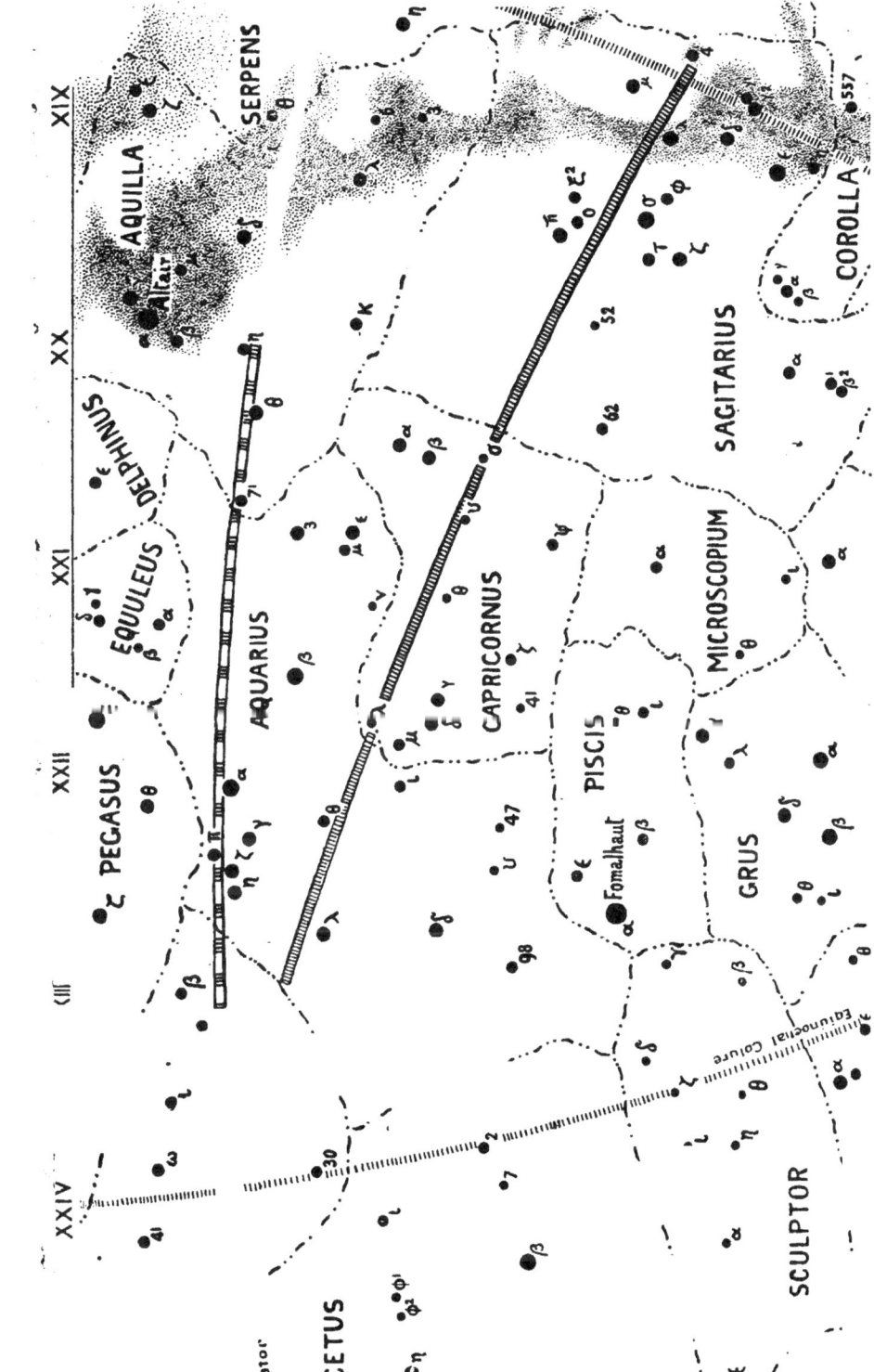

southern heavens are generally given over to creatures of like ill significance—Hydra, Scorpio, Lupus, Corvus, Canis—representing, with Cetus, and perhaps Lepus, the " Seven Evil Spirits " of Akkadian mythology.

To the eye, the principal stars of Cetus form the outline of a lounge chair, spreading over a vast expanse of sky, its length being 50°, its average breadth 25°. Though none of its stars are bright, Beta only being of the second magnitude, and Alpha about $2\frac{1}{2}$, the general outline can be pretty easily followed. Proceeding from east to west, Kappa, Alpha, Gamma, Delta and Omicron mark out the headrest of the lounge chair, or if we prefer so to speak of it, the lower jaw of the monster. The body of the chair is carried on by Theta and Eta, whence the axis of the constellation curves down to Beta, the footrest of the chair, or the tail of the beast. The back leg of the chair is marked by Epsilon and Pi, in a straight line with Omicron ; whilst Zeta, Tau and Upsilon, springing from Omicron in a graceful curve, mark the front leg. Only a few of the stars have distinctive names in common use at present. Alpha is known inappropriately as Menkar, the " Snout," as the title more strictly belongs to Lambda, the fifth magnitude star above it. Beta is Diphda, the "Frog," its full name being Diphda al Tania, the "Second Frog," the first being Fomalhaut. These two stars were grouped together by Aratus as well as by the Arabs.

> "The southern Fish beneath Aquarius glides,
> And upwards turns to Cetus' scaly sides.
> Rolls from Aquarius' vase a limpid stream
> Where numerous stars like sparkling bubbles gleam ;
> But two alone beyond the others shine ;
> This on the Fish's jaw—that on the Monster's spine."

The star of Cetus is Omicron ; its title of Mira, " Wonderful," being justly given to it because of its remarkable variations. It does not occur in Ptolemy's Catalogue, and the first recorded observation took place in 1596, when David Fabricius observed it as a third magnitude star. At its minimum it entirely disappears from the unaided sight, its brightness sinking down in the extreme case as low as magnitude $9\frac{1}{2}$. Its maximum is usually about that of the third magnitude, but it has been known almost to equal Aldebaran. Its period of variation occupies eleven months, for about half of which it is invisible to the naked eye. Under ordinary circumstances, therefore, its brightness at maximum is four hundred times that at minimum half a year earlier; whilst its extreme range of brightness is four times as great as this. Such a variation still baffles our every attempt to account for it satisfactorily; the explanation which obtains most currency, namely, that Mira is a dying sun, the surface of which from time to time is nearly hidden by enormous sunspots, is but a crude guess which leaves us still with many difficulties. It is certain that a variation in the case of our sun of but a tithe the amount of Mira's would leave this earth as bare of life as a meteorite, ere it had passed through a single course of its changes. Yet Mira is but the chief and representative of a large class of variable stars.

PISCES.

The idea of Cetus as the persecutor of Andromeda is carried out on our modern star atlases by the intervention of the zodiacal constellation of the Fishes, and indeed the arrangement is warranted by the description which Aratus has given us in his poem—

"Where the equator cuts the zodiac line
On the blue vault, the glittering Fishes shine.
Though far apart, a diamond studded chain,
Clasping their silver tails, unites the twain.

> The northern one more bright is seen to glide
> Beneath the uplifted arm and near the side
> Of fair Andromeda."

The knot of the riband actually rests on the neck of the Sea-Monster, and as the riband stretches northward to the Northern Fish, the latter is often represented as playing the part of Cetus, and actually fastening on the side of his devoted victim, who might otherwise have smiled securely at the distant hate of the Sea-Monster. There does not appear, however, to be any sufficient traditional authority for this relationship. The classical legends identify the twin fishes with Aphrodite and Eros, who plunged into the Euphrates to escape from the attack of the monster Typhon ; clearly only an imperfect Greek rendering of an Euphratean tradition. The doubling of the fishes is supposed by Sayce, but manifestly without due cause, to be due to the double month Adar, that is to say, Adar with the intercalary month Ve-adar. The constellations must have been, and were, mapped out long before the division of the zodiac into the twelve equal signs of 30° each, associated with the months.

Although Pisces is a particularly dull constellation to the naked eye, its two brightest stars, Eta and Gamma, barely surpassing the fourth magnitude, it is an easy constellation to trace out, as it consists chiefly of a number of stars between the fourth and fifth magnitudes. arranged in two obvious streams. Alpha, the most easterly star of the constellation, is not far from Mira Ceti, and marks the knot where the two cords attached to the tails of the two fishes are tied. This is implied by its name Okda, from Okda al Kaitain, the "knot of the two threads."

> "The silken bands that join the Fishes' tails
> Meet in a star upon the Monster's scales.
> Beneath Orion's foot Eridanus begins
> His winding course, and reaches Cetus' fins."

ERIDANUS.

The constellation of the Stream is a large but not specially well marked one. Aratus and Eratosthenes give it the name "Eridanus," and it seems extremely probable that the Akkadians identified it with the Euphrates as the Egyptians with the Nile. Brown considers that the name "Eridanus" may be a Turanian river name, meaning "strong river." The meaning of the placing of the river in this part of the heavens is not apparent at first sight, since the story of Perseus and Andromeda, with which the Sea-Monster is connected, gives no record of a stream or river, unless we consider that the constellation refers to the flood, to cause the abatement of which Andromeda was sacrificed. If this be so, then no doubt Eridanus stands for the Great Deep of the Primeval Chaos of which the Sea-Monster typified the indwelling principle. Proctor has seen in the constellations surrounding Ara a note of the Hebrew account of the great Deluge, and it is possible that in this flood of Eridanus which the Sea-Monster pours forth from his mouth, there may equally be a reminiscence of the same great flood which figures so largely in Babylonian tradition.*

Only two of the stars of the constellation bear special

---

* Or it may be connected with an Akkadian tradition referring to a yet earlier time; "when within the sea there was a stream" which flowed on both sides of Êridu, the Eden-city, situated in the Abyss.

"The Abyss had not been made, Êridu had not been constructed,
The glorious house, the house of the gods, its seat had not been made—
The whole of the lands were sea.
When within the sea there was a stream,
In that day Êridu was made."

Pinches' *Translation of the Akkadian Creation Legend.*

names in common use to-day—Alpha Eridani, a bright star far below our English horizon, is now known as Achernar, meaning " the end of the river." In Ptolemy's Catalogue the star with the corresponding title is our Theta, a star some 17° further north. But apparently it was felt the constellation lost its meaning unless it was prolonged downwards to the horizon, and hence as new stars came into view with southern exploration, the constellation was prolonged to within 32° of the southern pole. Beta Eridani, the brightest star in the part of the constellation familiar to us, and close to Rigel, the bright star on Orion's knee, is sometimes known as Cursa, the " footstool " or " throne " of Orion.

ARIES.

Close beneath the lower points of the greater W, noticed above in describing Perseus, close, that is to say, beneath Beta Persei and Beta Trianguli, is the small but distinguished constellation of the Ram. He can also be found readily by following the stars which mark the belt of Andromeda. Proceeding from Nu Andromedæ through Mu and Beta, we find at double the distance between Nu and Beta the extremely flat triangle which marks the Ram's head. As Aratus tells us :—

> " No splendid gems his golden fleece adorn,
> Two dimly glitter on his crooked horn.
> If you would find him in the crowded skies,
> Beneath Andromeda's bright belt he lies."

Of these three stars, the brightest, Alpha, is the furthest to the north and east; Beta, the next in brightness, is next also in position; and Gamma, the most southern and westerly, is the faintest. A compact little triangle about half-way from Alpha Trianguli to the Pleiades comprises practically all the other important stars of the constellation.

The mythology of the constellation is as little striking as is its actual appearance in the sky. It is, of course, associated in Greek mythology with the voyage of the Argo; it is indeed the Ram whose golden fleece the Argonauts went forth to seek. But the story carries no conviction with it, nor is there anything in the surrounding constellations to support the legend, although so great an astronomer as Sir Isaac Newton held that the stellar symbols were intended simply as a record of the Argonautic expedition.

An explanation which at first sight appears more plausible, asserts that the Ram was the constellation in which the sun was placed at the spring equinox at the time when the constellations were mapped out, and that as the first of the zodiacal twelve, it was natural to symbolize it as the bell-wether of the starry flocks. Unfortunately for this theory we know that the constellations were mapped out at a far earlier epoch, when the equinox fell in the middle of the constellation Taurus.* At that date the constellation of the Ram came last of the zodiacal twelve, and to represent it as the stellar bell-wether would have been absolutely the most unnatural thought possible.

Yet from an early age it has had that position. It is the leading sign in the systems of astrology which have come down to us through the Greeks, and it figures as the leading sign in most of the explanations of the

---

*Mr. R. Brown, on page 54 of Vol. I. of " Primitive Constellations," following an obvious slip of Prof. Sayce, says that the stellar Ram " onwards from B.C. 2540 opened the year," instead of "from 2540 *years ago*," the time when the equinox really fell amongst the three stars of the Ram's head.

constellation figures which have come down to us from antiquity.

There is a great significance in this fact. It proves at once that these astrological systems and these theories of the constellation figures, not only took their rise at a later epoch, but that when they took their rise the real origin and meaning of the designs had been wholly lost.

The only stars in the constellation that are usually known by their Arabic names, are the three in the head. Alpha is known as Hamal, the "Ram," the brightest star being put for the entire constellation. Brown indeed asserts that the stellar Ram was in the first place only the star Hamal, the constellation being formed round it afterwards. In Chaucer the star is referred to as Alnath, that is to say, "the horn-push," a name more commonly associated nowadays with the star on the tip of the northern horn of the Bull, Beta Tauri, to which it is even more appropriate.

> " And by his eighte speres in his working,
> He knew full well how far Alnath was shove
> Fro the head of thilke fix Aries above,
> That in the ninthe spere considered is."

Since the actual stars were accounted to be placed in the eighth sphere whilst the twelve equal signs of the Zodiac, each of thirty degrees of longitude, were placed in the ninth, Chaucer is here stating that his astrologer knew the distance from the first point of Aries, that is to say from celestial longitude 0°, to the first star in the actual constellation Aries. In other words, he knew the amount of precession.

Beta and Gamma bear the names Sheratan and Mesartin, the former word meaning "the two signs," the latter "the two attendants." The two names taken

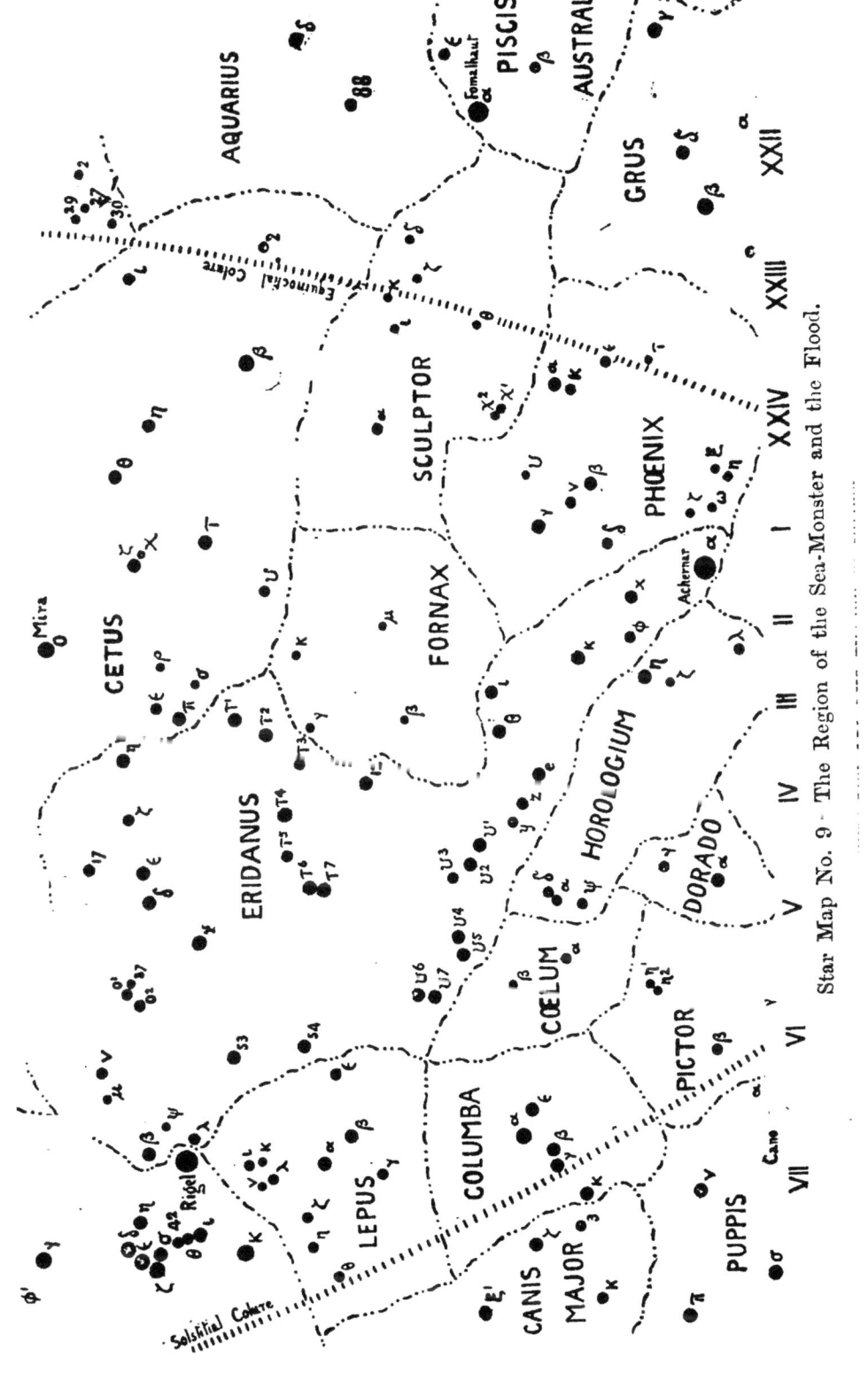

Star Map No. 9 - The Region of the Sea-Monster and the Flood.

together therefore mean " the two attendant signs," these two stars being considered as attendants on Hamal.

A small constellation in this region of the sky may receive a passing notice. Immediately below the Dolphin and close by the head of Pegasus, is the little group Equuleus, the " Forepart of a Horse." It is placed in the sky in such a manner as to suggest that Pegasus is not rushing through heaven alone, but is in close company with a yoke fellow, whose head, indeed, is just seen, but whose body is hidden from us by that of Pegasus. The constellation is not mentioned by Aratus, but it is stated that Hipparchus formed it from stars formerly belonging to the Dolphin. It is described as we now have it in the catalogue of Ptolemy.

CHAPTER VII.

THE STARS OF WINTER.

THE long nights of winter are the time when the heavenly hosts gather in their most resplendent squadrons. Sirius, by far the brightest of all the fixed stars, reaches the meridian at midnight of New Year's Day. Orion, the most splendid single constellation, is crossing from 10.30 to 11.10, the same night. Procyon, the Lesser Dog star, follows Sirius in its southing by about forty minutes. East and west of these are the two bright zodiacal constellations of the Twins and the Bull. South and east of Sirius is the hugest of the constellations, the ship Argo, resplendent with many brilliant stars, but distinguished amongst all the stellar groups by the numbers and the compact clustering of the small stars just clearly within the grasp of the unaided sight. Aratus marks the special glory of this region, though Dr. Lamb fails, as he too often does, to represent his exact meaning.

> "First rise athwart the Bull—majestic sight,
> Orion's giant limbs and shoulders bright,
> Who but admires him stalking through the sky,
> With diamond-studded belt and glittering thigh.
> Nor with less ardour, pressing on his back,
> The mottled Hound pursues his fiery track.
> Dark are his lower parts as wintry night,
> His head with burning star, intensely bright,
> Men call him Sirius, for his blasting breath,
> Dries mortals up in pestilence and death."

TAURUS.

There is no difficulty at all in finding the principal stars of the Bull, the original leader of the zodiacal twelve. The Pleiades, to the naked eye by far the most striking star cluster in the heavens, mark the shoulder of the Bull; the Hyades, a looser, but yet brighter group, mark its head. The latter group forms an exceedingly well-marked capital V. Aldebaran, Alpha Tauri, a bright orange star, marks the upper left hand of the V, Epsilon Tauri similarly marks the point on the right, whilst Gamma marks the angle. Theta 1 and Theta 2 lie between Alpha and Gamma, and three stars marked Delta lie between Gamma and Epsilon. Following the two branches of the V eastward to about four times their length, we come to Beta and Zeta, which mark the tips of the two horns, Beta, the northern horn, being much the brighter of the two. These are the chief stars in the constellation, for the Bull, like Pegasus, is shown only in half length, and beside his head and horns possesses little but the forelegs. Iota Tauri stands just mid-way between Beta and Zeta on the one side, and Alpha and Epsilon on the other. A straight line from Iota through Gamma leads to Lambda Tauri, and followed on, passes just below Omicron Tauri, which forms a pair with Xi Tauri.

The names in the constellation in familiar use to-day are restricted to the names for the two groups, the Pleiades and the Hyades, to the classical names for the individual stars in the former group, and to the two stars Alpha and Beta. Beta, as already noted, is Alnath, "the horn-push"; Alpha is Aldebaran, "the follower," that is to say, of the Pleiades.

" Near Perseus' knee, the Pleiads next are rolled
  Like seven pure brilliants set in ring of gold;
  Though each one small, their splendour all combine
  To form one gem, and gloriously they shine.
  Their number seven, though some men fondly say
  And poets feign that one has passed away.
  Alcyone—Celœno—Merope—
  Electra—Taygeta— and Sterope—
  With Maia,—honoured sisterhood—by Jove
  To rule the seasons placed in heaven above.
  Men mark them rising with the solar ray
  The harbinger of summer's brighter day.
  Men mark them rising with Sol's setting light
  Forerunners of the winter's gloomy night.
  They guide the ploughman to the mellow land;
  The sower casts his seed at their command."

Of all the star groups none figures so largely in myth
and legend and literature as the Pleiades. The name
Pleiades is probably derived from the Greek Pleiones,
" many," and accords with the Hebrew *Kimah*, "the
cluster"; the Babylonian "Kimtu," "the family," and the
Arabic Ath-thurayya, " the little ones." The names given
to the individual stars are those of the seven daughters
of Atlas and Pleione. According to Aeschylus they
were placed in heaven on account of their filial sorrow
when their father was turned into the mountain and
laden with the weight of the firmament. Of the
numerous other names which have been given them, the
" Hen and Chickens " is one of the most familiar. Thus
Miles Coverdale in a marginal note to the book of Job
in his translation of the Bible, calls them "the Clock
Henne with her chickens." Many of the Greek poets
refer to them as "the doves " or "rock pigeons,"
writing their name as " Peleiades." They were likewise
the " Vergiliae," "the stars of spring," or the " Atlan-
tides," from their father. Modern astronomers have

commemorated Atlas and Pleione, the parents of the seven, in the pair of stars on the east of the group.

Their value to the ancients as an indication of the seasons and the progress of the year was immense. As noted in Dr. Lamb's unduly expanded paraphrase from Aratus quoted above, their heliacal rising showed the commencement of summer, their acronical rising the

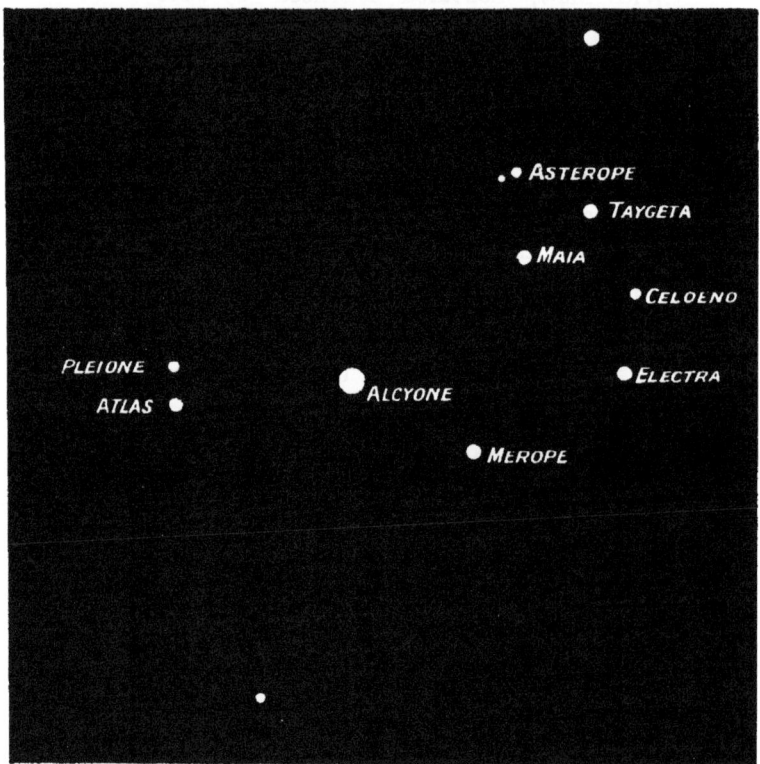

Fig. 21.—The Pleiades.

commencement of winter. Four thousand years ago they marked the position of the sun at the spring equinox, and were therefore then especially closely connected with the revival of the forces of nature in the opening year. This no doubt is the first meaning of the reference to " the sweet influence of the Pleiades" in Job xxxviii. Whilst in conjunction with the sun they were hidden from view for forty days, reappearing as summer drew near. Thus Hesiod sings:—

> "There is a time when forty days they lie
> And forty nights concealed from human eye;
> But in the course of the revolving year
> When the swain sharps the scythe, again appear."

In November, on the other hand, they are up all night, and are on the meridian at midnight. In some mysterious way they became associated with the memorial observances for the dead which in so many nations have always been associated with this month of November, and which seem to point to the great Christian commemoration of All Saints' Day.

A keen sight will count nine or ten stars in the group. Moestlin, according to his pupil Kepler, counted fourteen. Miss Christabel Airy plotted twelve; but these high numbers are hardly obtained legitimately, inasmuch as they include stars lying at some considerable distance from the main portion of the group. Six stars are easily seen, and amongst nations the most widely scattered and unconnected, we find the tradition that though six stars are only visible to-day, seven was their original number. The probability seems that this is no mere legend, but the record of an astronomical fact. Several of the present Pleiades are slightly variable, and Alcyone, the leader in brightness to-day, would seem to have only attained that rank within the last few centuries. It is therefore quite conceivable that one of the members of the group, now too faint to be detected without a telescope, may in time past have fully rivalled her sisters.

Even such small magnifying power as an opera-glass can lend greatly increases the beauty and interest of both the Pleiades and the Hyades, and the whole region between the Bull's horns. Close to Zeta Tauri, the southern horn tip, is the Crab nebula, the first in Messier's catalogue, and, from its comet-like appearance, a snare to beginners in the art of comet seeking. Zeta Tauri marks very nearly the radiant point of a shower of very slow and bright meteors, active at the end of November and early in December; whilst Epsilon Tauri is not far from the radiant point of the other great Taurid stream, active from October 20 to the end of November.

Lambda Tauri is a variable star of the same class as Algol, Beta Persei; that is to say, its variation is due to the transit across it of a dark companion. Its period is 1h. 12m. short of four days; its maximum, or rather ordinary light, ranks it of magnitude 3·4; at minimum it sinks down to 4·2.

AURIGA.

"Next the broad back and sinewy limbs appear
Of famed Auriga, dauntless charioteer,
Far in the north his giant form begins
Reaching athwart the sky the distant Twins,
The sacred goat upon his shoulder rests,
To infant Jove she gave a mother's breasts.
Kind foster nurse! Grateful he placed her here,
And bade her kids their mother's honour share.
  *   *   *   *   *   *   *
Auriga and the Bull together meet,
Touches his star-tipped horn the hero's feet.
The Bull before him to the west descends
Together with him from the east ascends."

Beta Tauri, the northern horn, belongs also traditionally to the constellation of Auriga; now known as "the charioteer," although no chariot is visible, and in Ptolemy's star list as well as in our modern repre-

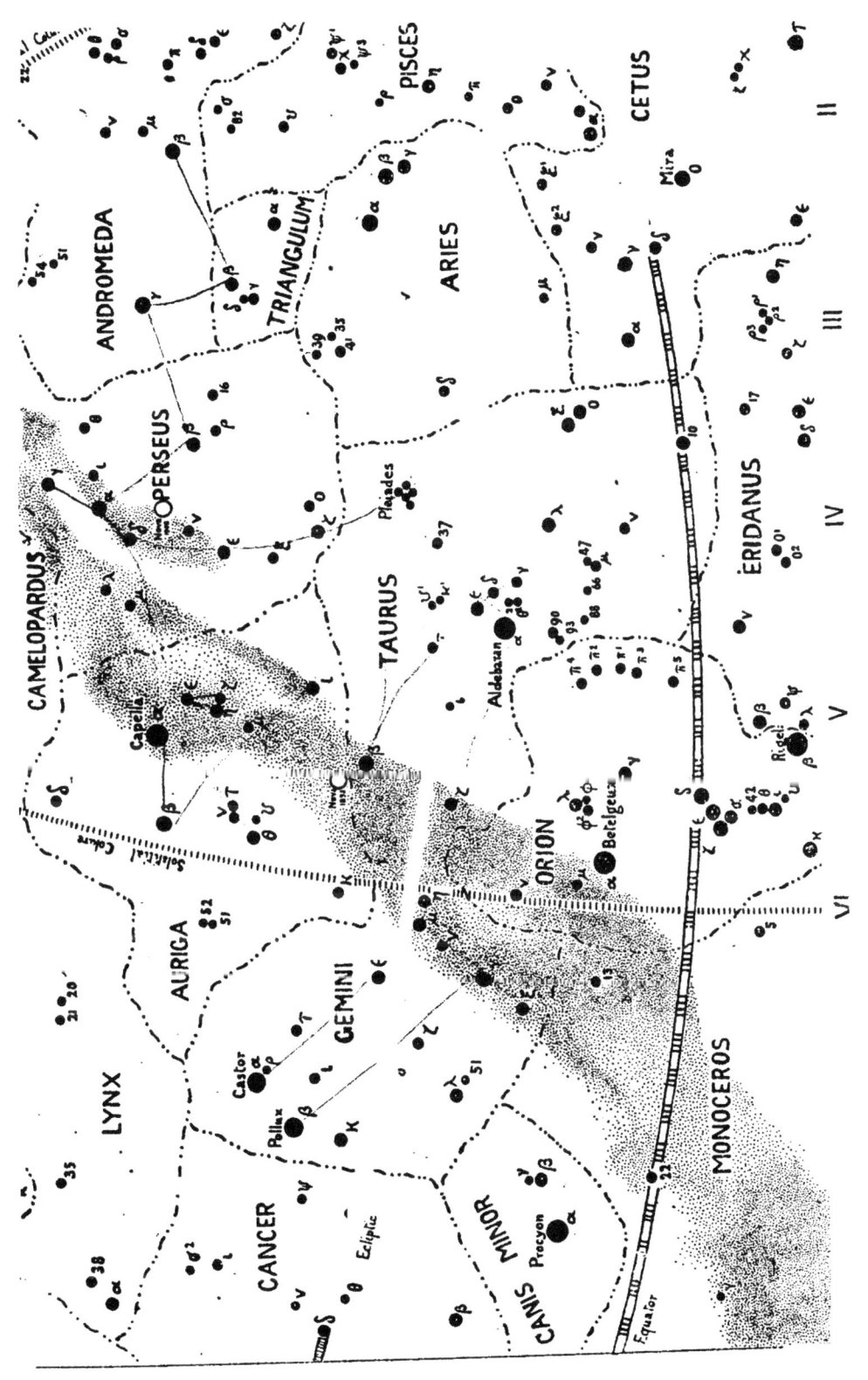

sentations he is described as in the attitude of a
shepherd, carrying a goat on his shoulder and a pair of
little kids in his hand.   The Greek name for the constellation is Heniochus, "the holder of the reins"; a
name preserved for us in the Arabic name for Beta
Aurigæ, "Menkalinan," "the shoulder of the rein-
holder."

The chief stars of the constellation are easily picked
out.   Capella, a bright star of a yellowish creamy light,
nearly balances the steel-blue gem Vega on the opposite
side of the pole star, and with Menkalinan (Beta Aurigæ)
has already been described as one of the great guide stars
of the circumpolar regions, and the zenith star for the winter

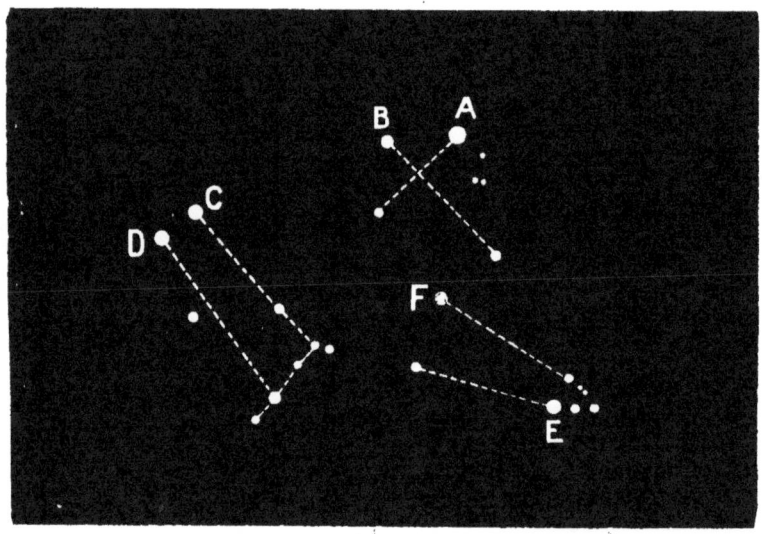

A, Capella.   B, Menkalinan.   C, Castor.   D, Pollux.   E, Aldebaran.
F, Alnath.
Fɪɢ. 22.—Auriga, Gemini and Taurus.

midnight.   Taking Beta as the head star of a cross, Capella
marks the western extremity of the crossbeam, Theta the
eastern, and Iota the foot.   A straight line from Theta upwards through Beta leads to Delta, a star which marks the

head of the figure. Of the three stars marking the little triangle by Capella, Epsilon is nearest to Capella, and Eta and Theta, known as the Haedi, "the kids," mark the short side of the triangle.

In recent years the constellation has been most distinguished by the appearance of the new star discovered on February 1, 1892, by Dr. T. D. Anderson, using only a small pocket telescope. The position of the Nova is in the extreme south of the constellation, about 3° to the north of Beta Tauri. At the time of discovery it was about the 5th magnitude, and slightly brighter than Chi Aurigæ, which is just 2° to the north of it.

GEMINI.

Passing on further to the west, a pair of bright stars are seen as far below the forefoot of the Great Bear as Alpha is above it. These are Castor and Pollux, the two chief stars of the constellation Gemini, the "Twins." This constellation is, according to Brown, a stellar representation of the great Twin Brethren of the sky, the sun and the moon, who join in building a mysterious city and who are hostile to each other although they work together. In classical legend they are the children of Zeus and Leda, the Dioskouroi, and by no means have the fratricidal relation which this interpretation would suggest. The idea of strife between sun and moon is natural enough, and no doubt many stories like that of Romulus and Remus took their form from such a nature myth. But the idea of strife is not the leading one in most of the legends relating to the stars Castor and Pollux, who are indeed shown as man and woman on many Zodiacs, and I think that this fact renders it questionable if it is safe to press far the theory that the ancients, so to speak, solarized the stars, designing the constellations to perpetuate the stories in which they had dramatized their conceptions of solar and lunar relations.

The limits of the constellation are easy to trace out. Four fairly bright stars, nearly in a straight line, mark the feet of the Twins, whilst Castor and Pollux mark their heads. They are standing in fact upon the Milky Way. The four feet stars are Mu, Nu, Gamma, and Xi. Close by Mu is Eta, another star in the foot of Castor, and in the neighbourhood of these two stars is a splendid cluster, 35 Messier, which is just visible to the naked eye, but which well repays examination with an opera-glass. Castor is one of the most celebrated of double stars, though of course altogether beyond the grasp of an opera-glass as the components are not 6″ apart, but the constellation as a whole is a fine one for examination with the opera-glass, especially in the region of the Milky Way. The contrast in colour between the two principal stars, Castor and Pollux, is noticeable enough even with the naked eye, but becomes very striking when the glass is turned upon them. Gemini forms the home of several important meteor-radiants, the principal one being that of December 10 to 12.

Zeta Geminorum, which marks the knee of Pollux, is an opera-glass double, a seventh magnitude companion being distant from it a minute and a half.

ORION.

A straight line dropped from Capella through Beta Tauri, the northern horn of the Bull, leads straight to the centre of Orion, the brightest and most famous constellation in the sky, and the most frequently alluded to in literature, ancient or modern, sacred or classical.

Both the Authorised and the Revised Versions of the Bible refer the *Kesil* of Job xxxviii., 31, to this constellation. There is much probability that the rendering of the Revised Version for the other two constellation names mentioned in this text, *Aish* and *Kimah*, "the Bear" and "Pleiades," are quite correct, but there is more uncertainty in the present identification. *Kesil* means "impious, mad, rebellious," and as such is traditionally supposed to refer to Nimrod, "the mighty hunter before the Lord," supposed to be the first great conqueror, and the first to set up a tyranny based upon military power. One difficulty in rendering *Kesil* by Orion is that the same word occurs in the plural in Isaiah xiii., 10, where the word is translated "constellations." If *Kesil*, therefore, really refers to Orion, we must suppose that in this passage the most glorious constellation of the sky is put for constellations in general; unless, perchance, Bootes was regarded as a second Orion, as suggested on page 52. The context, however, would rather lead to the idea that we should look for a winter constellation to correspond to *Kesil*; for just as "the sweet influences of the Pleiades" evidently refer to the revival of nature in the spring, so "the bands of Orion" may be naturally supposed to point to its imprisonment by the cold of winter. If Nimrod be really the original Orion, there was an unsuspected appropriateness in the sycophantish proposal of the University of Leipsig to give the centre stars of the group the name of Napoleon, the most modern example of the same mad ambition.

Brown traces the name Orion to the Akkadian Ur-ana, "the light of Heaven," a poetical and most natural title for the most beautiful and brightest of all the stellar groups. And it may well have been that, as Brown further thinks, this name was given because the constellation was taken as a stellar reduplication of the one great light of heaven, the sun; or the same name having been given independently to both the sun and the constellation, the latter was taken as representing the former. The stellar giant, therefore, on this view presents to us a personification of the sun, " rejoicing as a giant to run his race."

But what does he pursue? His prey is found in the little constellation beneath his feet, one of no distinction or brilliancy, the Hare.

" Up from the east the Hare before him flies,
   Close he pursues her through the southern skies."

Now it is certain, as Brown points out, that " the amount of folklore and zoological myth which all over the world connects the moon and the hare is simply astonishing." Of course it does not necessarily follow that the Hare as a constellation is also a symbol of the moon, but at least the suggestion has no improbability. It is possible, therefore, that in Orion the mighty hunter trampling on the timid fleeting hare, we ought to recognise a primeval emblem of the rising sun overpowering and crushing with its vastly more powerful light the feebler rays of the moon as she flees before him towards the west. But whether the two Dogs which we find attending and following Orion have any deeper meaning than the natural desire to piece out this picture of a hunter and his chase, by providing him with a leash of hounds, may well be doubted.

The figure of the giant hunter is one of the very easiest to make out of all the constellation figures. Seven bright stars stand out with special distinctness. That furthest to the north-east of the seven is obviously orange in colour, and is Alpha, Betelgeuse, " the shoulder of the giant." The star in the north-western corner is Gamma, Bellatrix, "the female warrior." This last title is from the translation in the Alphonsine Tables of the Arabic title Al Najid, "the conqueror." The south-western corner is held by Beta, Rigel, the "foot," the brightest star of the constellation. The fourth corner, the south-eastern, belongs to Kappa, now known as Saiph, "the sword," the name having been transferred to this star from Iota, to which it really belongs. Three stars mark the Belt of the giant, as the four foregoing mark his two shoulders and his legs. These in succession are Delta, Mintaka, "the girdle"; Epsilon, Al Nilam, "the string of pearls"; and Zeta, Al Nitak, "the belt." The sword is marked by a short row of stars in a straight line below Epsilon. These are, 42, Theta and Iota. To the eye, Theta is a misty star; its diffused appearance being due to the great nebula, the most glorious object in the heavens. Between Alpha and Gamma, but a little to the north, is a compact little triangle of stars, Lambda and Phi 1 and Phi 2, which mark Orion's head. His club stretches up across the Milky Way to the feet of Gemini and the horns of Taurus, whilst between Bellatrix and Aldebaran a curving line of stars runs nearly due south, marking the lion's mane, which the Hunter is shaking before the eyes of the Bull, as if it was the scarlet cloak of a toreador.

LEPUS.

The little constellation of the Hare does not contain much of interest for the naked-eye observer. Its principal star, Alpha, sometimes known as Arneb, "the hare," is very near the point of an equilateral triangle, of which Beta and Kappa Orionis form the base.

CANIS MAJOR AND CANIS MINOR.

A far more majestic triangle, and more nearly exact
in its proportions, is that formed by Betelgeuse, Sirius
and Procyon.

"Let Procyon join to Betelgeuse and pass a line afar,
    To reach the point where Sirius glows, the most con-
        spicuous star;
    Then will the eye delighted view a figure fine and vast;
    Its span is equilateral, triangular its cast."

The constellation of Canis Major, though in itself a
brilliant one, lies so low for English observers that
practically we think of little but its chief star. But
this is so far and away the brightest in the sky, being
more than two full magnitudes brighter than Aldebaran
or Altair, average first magnitude stars, and is rendered

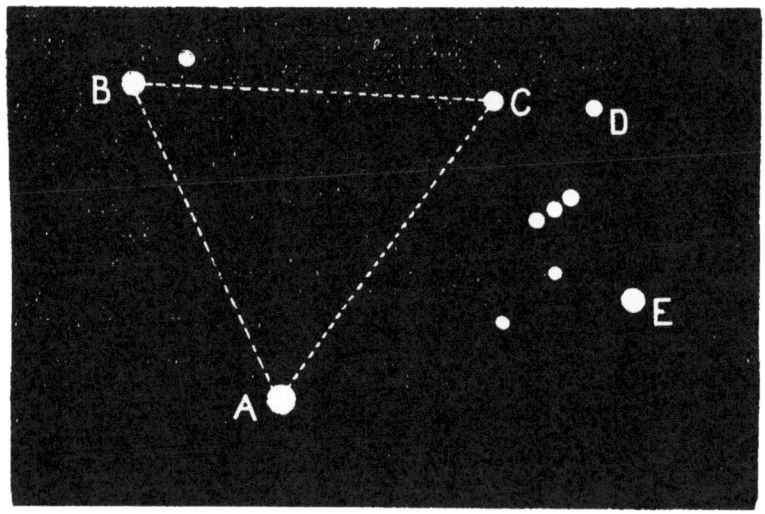

A, Sirius. B, Procyon. C, Betelgeuse. D, Bellatrix. E, Rigel.

FIG. 23.—Sirius, Procyon and Orion.

so unusually striking by the intensity of its scintillations,
that it serves alone to amply distinguish the constellation
to which it belongs. Its excessive scintillation is due

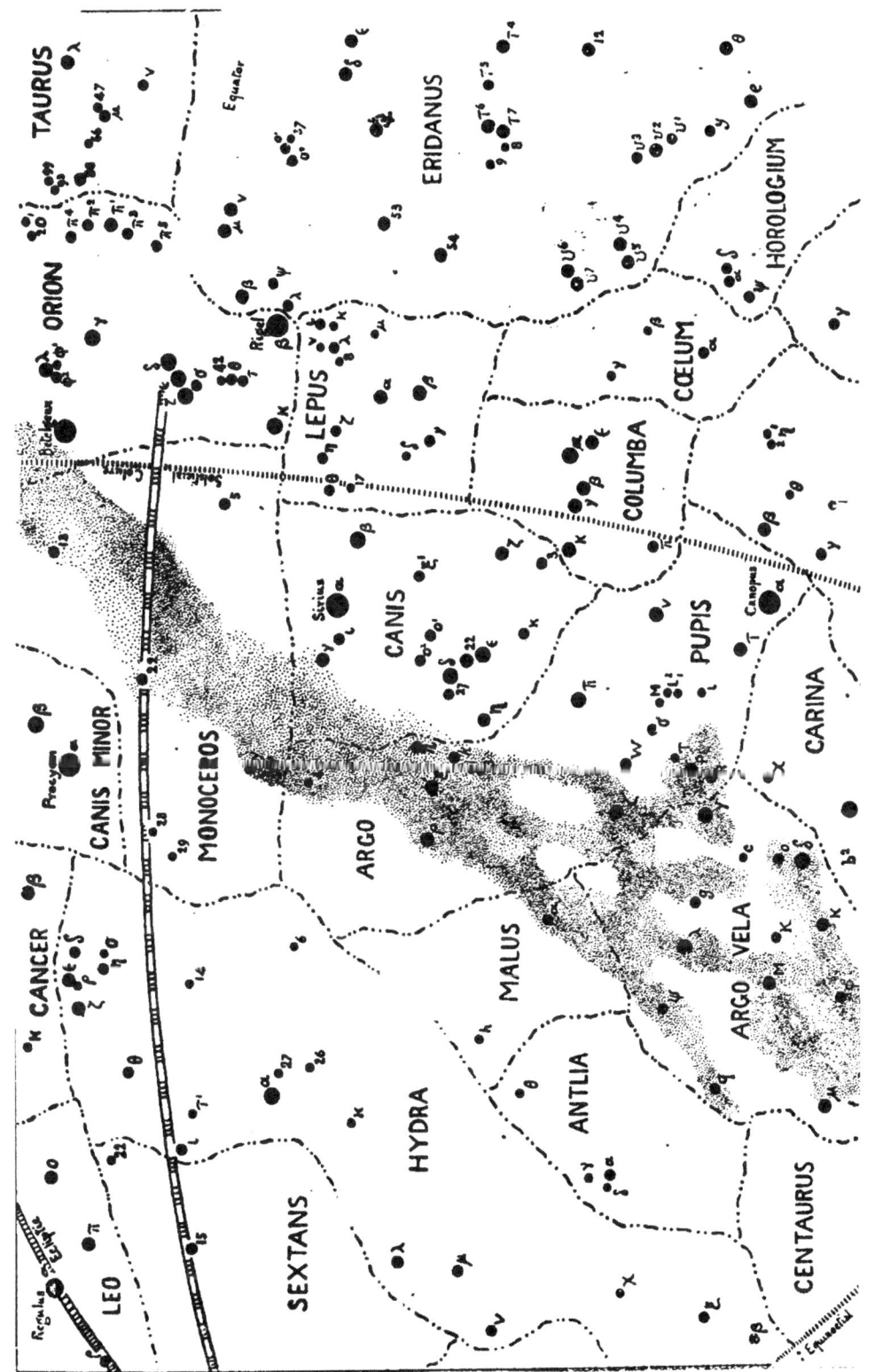

to two causes; the one that in these latitudes it can never attain any great elevation; the other, its striking bluish-white colour. For stars of that colour scintillate more markedly than those of any other, and Sirius being the very type and model of the class, its flashing renders it as conspicuous as its brightness.

> "The fiery Sirius alters hue,
> And bickers into red and emerald."

The name Sirius is usually taken as bearing its Greek meaning of "sparkling," "burning," or "scorching," an appropriate enough name for the brightest of the stars and one in conjunction with the sun at the beginning of summer. Other renderings interpret it as "chief," or "prince," or "the bright and shining one"; all equally natural and appropriate for this prince of stars. Our name for the Lesser Dog, Procyon, is simple enough. It is merely "before the dog"; in other words, Procyon was the forerunner of Sirius, not as crossing the meridian earlier, but as rising before it and so heralding its appearance.

The colours of the four brightest stars of this region are very well worth studying; the orange of Betelgeuse being in strong contrast with the steel-blue of Rigel and Sirius, whilst Procyon, though a white star, yet shows a distinct creamy or yellowish tinge, which to a sensitive eye is on a fine night very clearly distinguishable from the colder hue of the two brilliants first named.

The constellation of the Lesser Dog in Ptolemy's catalogue numbers only one star in addition to Procyon. This is Beta, Gomeisa the "dim-eyed," possibly as meaning that Procyon is so much less brilliant than Sirius. Or it may be that it is a corruption of Al Gamus, the "puppy." A glance at the sky shows that Procyon and Gomeisa stand at much the same distance apart as Castor

and Pollux, and the Arabian astronomers used to call
the space between the two stars of the Twins " the long
cubit," whilst that between the two of the Lesser Dog
was " the short cubit,"—an old instance of the tendency,
which seems so irresistible, to account a space in the
heavens of some seven to ten degrees of arc in length
as equivalent to a terrestrial yard.

Procyon and its companion form yet another proof
that the constellation names were not given in couse-
quence of forms suggested by the actual groupings of
the stars.   Certainly there is nothing to suggest a dog
in the presence of two fairly bright stars some five
degrees apart.

The Greater Dog is a far fuller constellation.   Beta
precedes Sirius some twenty-two minutes and is therefore
called Murzim, the " Announcer."   Quite low down near
our horizon is a right-angled triangle of bright stars,
the right angle being marked by Delta.   Epsilon and
Eta at the two other points of the triangle lie some
three degrees further south and are nearly on the same
parallel with each other ; Epsilon is Adhara, the " back."
Delta is in the centre of a very interesting region, a
curious curve of small stars preceding it, whilst another
follows.

Between the two Dogs and following Orion, is a large
constellation without any conspicuous stars which was
formed by Bartschius, the son-in-law of Kepler.   This is
Monoceros, the Unicorn, and it marks a region rich in
double stars, star clusters and nebulæ, but without any
attractions to the naked-eye astronomer, unless we except
the star S Monocerotis, which is variable within a range of
half a magnitude, and in a period of about three and a
half days.

CHAPTER VIII.

The South Circumpolar Stars.

THERE is a strange, unforgettable sensation in the first voyage from our high northern latitudes to the southern hemisphere. Night after night the old familiar stars, the constant occupants of our sky at home, sink lower and lower in the north and disappear, whilst new lights rise to shine upon us from the south. But beside the disappearance of old friends and the coming into sight of stranger stars, the known stars that still remain to us adopt most unfamiliar attitudes, and these become more and more perplexing the further south we go. Lordly Orion stands on his head in the exact attitude of a little city Arab, turning a cart-wheel; the long procession of the Zodiac, all in turn suffer the same inversion; Pegasus alone, a topsy-turvy constellation with us, pursues his course across the sky with head upraised. Even the moon seems altered from the moon we knew in Britain; the mournful face that watched us there is no longer recognisable as such, since we see "his spotty globe" inverted.

The new regions of the sky opened to us from southern latitudes not only differ in detail from those with which we are familiar, they possess some general characteristics that cannot fail to strike the observer. Starting from Scorpio, here no longer creeping along the horizon as in England, but riding in the very zenith, we find the Milky Way sweeping downwards towards the south-west in masses of the greatest brilliancy and complication. Around its course is gathered the greatest aggregation of brilliant stars to be found anywhere in the sky. The Scorpion is followed by the Wolf and the Centaur, and these by the ship Argo. Between Argo and Centaurus comes the little Southern Cross, a constellation which has perhaps been extravagantly praised, but which yet contains within a very small area three stars above the second magnitude, and six others ranging between that and the fifth. The Centaur claims ten stars brighter than the third magnitude, the Wolf three, Argo fifteen, one of which, Canopus, is inferior only to Sirius. As the Milky Way sweeps through Argo, it enters Canis Major, which, though not a large constellation, contains Sirius and three stars above the second magnitude and three more above the third. This belt of the sky, therefore, from Scorpio to Sirius, is by far the most brilliant in the entire heavens, both from the numbers of brilliant stars clustered within a narrow band and from the brightness of the Milky Way. But beyond this belt we find an altogether different state of things. Round the southern pole there are no star groups to rival the Great Bear, Cassiopeia, the Dragon, or even Cepheus and the Lesser Bear, which surround the northern pole. The new constellations, which were placed round the southern pole by Dircksz Keyser, and Lacaille, are for the most part entirely destitute of bright stars, or of easily remembered configurations; those actually round the pole being especially poor in this respect. Octans, the constellation in which the southern pole is situated, though of considerable extent, contains but a single star as bright as the fourth magnitude, nearly all its members being fainter than magnitude $5\frac{1}{2}$. The southern pole star, Sigma Octantis, is catalogued as of magnitude 5·8.

The other chief features special to the southern heavens, are the presence of the two Magellanic Clouds, to which there is nothing to correspond in the northern sky. Then close to the Southern Cross, we find the Coal Sack, a great hole in the Milky Way, blacker and much more defined than any corresponding gap in its northern course. There is indeed a " coal sack " in the constellation of the Swan, but the portion of the Galaxy surrounding it is far from being so bright and distinct as that which hems in this curious island in the great celestial river of light. Lastly, the southern heavens generally seem much richer in stars just within the ordinary range of vision than the northern, the barren circumpolar region being almost surrounded by a broad belt, especially rich in the regions of Eridanus and Argo, in which sixth magnitude stars are to be found in unexampled abundance.

For Cape Town and the colonies of Victoria and New Zealand, the Southern Cross is a circumpolar constellation, always above the horizon, and at midnight on March 28th it stands vertically above the south pole, Canopus being then on the same level as the pole and west of it, whilst Alpha Eridani, Achernar, "the end of the river," almost touches the horizon in the south.

These great leader stars enable the general trend of the southern constellations to be easily made out. When the Cross is seen erect, the foot of the Cross is marked by Alpha, the head by Gamma, and the cross-beam by Delta and Beta, Beta being much the brighter of the two. Moving from Delta through Beta, we come at no great distance to the first magnitude star, Beta

Centauri, and Alpha Centauri lies but a little further on, almost on the same straight line. Nearly midway between Canopus and Alpha Crucis lies Beta Argûs, a star but little below the first magnitude. Not far from Beta is the False Cross, four stars in a trapezium, not quite so bright and a little more widely spread than those of the True Cross. Epsilon Argûs forms the foot, Kappa the head, and Delta and Iota the cross-beam.

Argo is scarcely at all an English constellation, the greater part of its huge hulk lying below our horizon. The vessel is drawn in our atlases and described by Aratus as travelling stern foremost. Thus in Brown's translation we read:—

> "Stern forward Argo by the Great Dog's tail
> Is drawn; for her's is not a usual course,
> But backward turned she comes, as vessels do
> When sailors have transposed the crooked stern
> On entering harbour; all the ship reverse,
> And gliding backward on the beach it grounds,
> Stern forward thus is Jason's Argo drawn."

The Greek legend of the voyage of the Argonauts under Jason to recover the Golden Fleece was believed by Sir Isaac Newton to have an actual historical basis, and to record the beginning of Greek commerce. In fact, in his view, the constellations generally were designed by the Greeks to celebrate the heroes and deeds of this great expedition. The knowledge which later times have brought us has compelled us to recognise that the constellations are far older than Newton thought them, and beyond a doubt Proctor was quite correct in recognising in Argo and the southern constellations near it—Centaurus, who is represented as having apparently just left the ship, and the Altar at which he is sacrificing—a pictorial record of that great Deluge of which the Hebrew and the Babylonian accounts agree so strikingly. The Altar lies entirely below our English horizon; the Wolf, the somewhat inappropriate animal whom the Centaur is offering up, shows but two or three faint stars; and Theta Centauri, the star in the Centaur's head, is the only bright member of that constellation visible to us.

Argo is so vast a constellation that for convenience of reference it has often been subdivided into four minor groups—Malus, the mast; Vela, the sails; Puppis, the stern; and Carina, the keel.

Generally speaking, the modern constellations, and particularly those in the southern hemisphere, have been conceived in the worst possible taste; the figures selected are utterly inappropriate to their surroundings, and unsuitable as means for identifying the stars. There is, however, one exception. If we take Argo as representing Noah's Ark, then Royer was unusually successful when he gave the name of Columba, Noah's Dove, to the group which he formed immediately south of Lepus. For, placed here—the group may be readily found when both Canopus and Rigel are seen, as the straight line joining the two passes through the constellation—the Dove appears to be perching upon the stern of Argo. Alpha Columbæ, the principal star, is sometimes known as Phact, "the thigh"; the thigh, that is, of the Greater Dog, whose territory has been invaded in order to give the greater room for the new group.

Those who have the opportunity of observing from southern or equatorial countries, will find the brightest and most easily recognised of all the new constellations due to Dircksz Keyser, just south of Piscis Australis, the constellation of which Fomalhaut is the chief star. This is Grus, the Crane, and five bright stars distributed

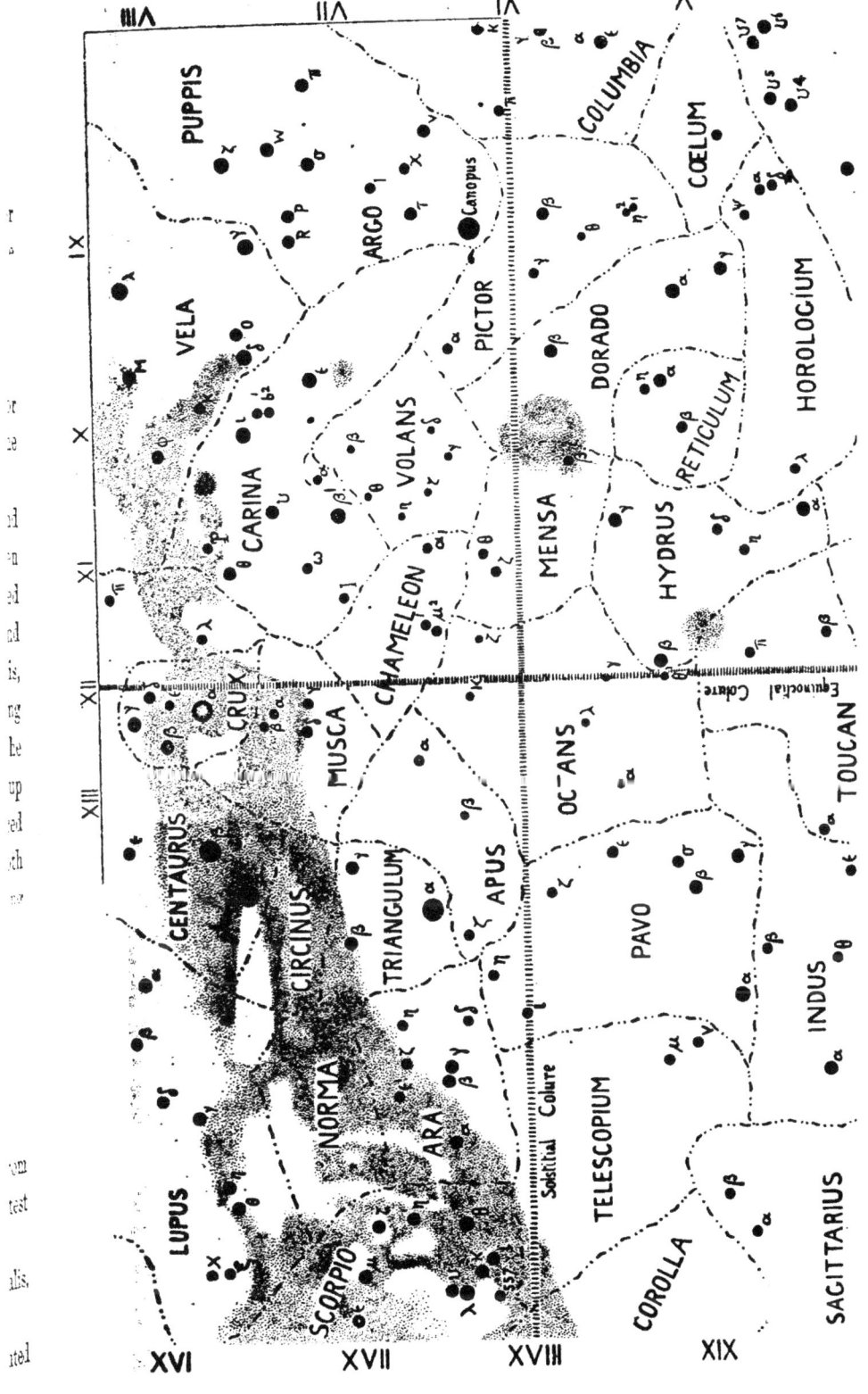

at equal distances along a gentle and regular curve, whilst a bright second magnitude star precedes them, form its principal characteristics. The Microscope, interpolated by Lacaille between Sagittarius and Piscis Australis, and the Sculptor's Tools, now more generally known as the Sculptor, which he added below Aquarius, and Cetus ; Fornax, the Chemical Furnace, which follows Sculptor; Antlia, the Air-pump, wedged in below Hydra and above the sails of Argo ; Cælum, the Sculptor's Graving-tool, below Lepus, and alongside Columba ; and Telescopium, which follows the Altar and lies beneath the Southern Crown, are as destitute of features of interest as the designs themselves are of appropriateness to their celestial surroundings.

The many obscure constellations which have been formed in the south circumpolar region may now be identified.

Between Octans, which surrounds the south pole, and the Southern Cross, we have Chamæleon and Musca; close to Beta Argûs is Volans, the Flying Fish; Pictor lies by the side of Canopus; the Greater Magellanic Cloud borders on four constellations, Mensa for "Table Mountain," Dorado, Reticulum and Hydrus; the Lesser Cloud lies between Hydrus and Toucan. Five constellations meet round Achernar; Eridanus, a primitive constellation; Horologium and Phœnix; and the two just mentioned, Hydrus and Toucan. Passing over to the Southern Cross, the Milky Way between that constellation and the Scorpion sweeps over the feet of the Centaur and six other constellations beside,—two of them primitive. The three constellations immediately below the Scorpion are Lupus, Norma and Ara, of which only Norma, placed in the middle of the Milky Way, is new. Circinus below Norma, and the Southern

Triangle below Ara are both new, as also is Apus, which lies between them and Octans. The Southern Triangle is a neat little figure on the edge of the Milky Way, its three principal stars making a diamond with Alpha Centauri. Alpha is the brightest star of the Triangle and the one furthest from the Centaur, lying midway between Alpha Crucis and Alpha Pavonis. The latter is a second magnitude star and the brightest in Pavo, the Peacock, the most brilliant of the new constellations, except Grus and Crux, and covering a considerable portion of the zone of small stars to which allusion has already been made. Indus, a fainter constellation, lies between Pavo and Toucan, and completes the band of small asterisms surrounding Octans.

SECTION II.

ASTRONOMICAL EXERCISES WITHOUT A TELESCOPE.

ASTRONOMICAL EXERCISES WITHOUT A TELESCOPE.

CHAPTER I.

THE SUN AND THE SEASONS.

It may seem at first sight a useless and idle suggestion that beginners in Astronomy should set themselves to the redetermination, on the roughest scale and with the simplest of instruments, of astronomical constants which were first determined more than five millenia ago, and which are now ascertained in our modern observatories to an almost inconceivable degree of exactness. Yet if we think for a moment we shall see that this is but the method which experience has taught us is the most effective in learning the other physical sciences. We know perfectly well we can never make a chemist of a boy by giving him a course of chemical text-books. We set him to repeat for himself experiments which were first made in the very infancy of the science. We make him determine again the combining weights of different elements, though these are known far more exactly than he can possibly work them out; and in so doing, he not only acquires skill as a worker, but the subjects of his study become real to him; he learns to know them in a sense which no amount of reading about them could ever supply.

It has been the drawback of Astronomy that this course has so seldom been adopted, and the inevitable result has been seen in that no science whatsoever has produced so large a proportion of paradoxers and cranks. There is no science the chief facts of which are so widely disseminated; there is none of which those facts are so little known by practical personal observation.

Much of this unfortunate state of things is due simply to the modern tendency to live in towns. Here the smoky atmosphere dulls the shining of the heavenly bodies; the crowded buildings hide the horizon and curtail the view of the sky, and at night tho artificial lights in streets and houses completely drown the feebler glitter of the stars and draw off attention from them. We do not need moon and stars as our ancestors did, and therefore we do not notice them. We, do not need to observe the sun to give us the time of the year; our almanacs tell us that. Therefore, except in observatories, the sun's place in the heavens remains unnoted.

But in early times this observation was of the very first importance. The constellations were mapped out some 5000 years ago, but before that was done—how long before we cannot tell—the length of the year had been determined and the apparent path of the sun amongst the stars had been laid down. The exact methods and instruments those early astronomers em-

ployed are not recorded, nor, if they were, would there
be any reason for slavishly copying them in repeating the
work to-day.   But in all probability the first astronomical
instrument was one of Nature's own providing, the natural
horizon.   And wherever a fairly good one is available, the
beginner in astronomy is strongly recommended to make
use of it.

If this were so then no doubt those primeval observers
had their attention drawn to the fact that as seen from
some given station the sun rose and set behind different
portions of the horizon at different times of the year.
In an open country free from mists and ground fogs,
this observation would be one of greater delicacy than
might be expected—a delicacy the greater according to
the distance of the horizon and to the number and
distinctness of the objects which could be recognized
upon it.   They would serve the purpose—so to speak—
of the divisions of a gigantic azimuth circle, and a few
years' careful and sedulous record of the exact position of
the sun at rising and setting would give an exceedingly
close determination of the true length of the tropical
year.

They would do more than this.   They would give the
means of determining the south point of the horizon—in
other words of the meridian line.   A line drawn at right
angles to the line joining the point of rising and setting,
would be *roughly* but not precisely the meridian line.   But
the mean of all the points thus indicated as due south
would, unless the horizon were much more obstructed on
one side than the other, approximate very closely indeed
to the true south point.

The conditions for different observers will vary so
widely that it would be useless to give detailed directions
as to making this observation, and it would be useless

for another reason. It is most important that those who take up the pursuit of naked-eye Astronomy should make their observations independently, and too detailed instruction beforehand would defeat the very object for which those observations were made.

It would soon be felt that the natural horizon was a rough and inconvenient instrument to work with. The objects ranged along it which serve as division marks are apt to be irregular, the horizon itself to deviate very considerably from an ideal plane. So perhaps the next step in the observation of the sun would be the erection of some means of observing the shadow it casts —in other words a simple sun-dial.

It is probable that the earliest sun-dial was simply the spear of some nomad chief, stuck upright in the ground before his tent. Amongst those desert wanderers, keen to observe their surroundings, it would not be a difficult thing to notice that the shadow shortened as the sun rose higher in the sky, and that the shortest shadow always pointed in the same direction—north. The recognition would have followed very soon that this noon-day shadow changed in its length from day to day. A six-foot spear would give a shadow at noonday in latitude 40° of 12 feet at one time of the year, of less than 2 feet at another. This instrument, so simple, so easily carried, so easily set up, may well have begun the scientific study of Astronomy, for it lent itself to measurement, and science is measurement; and probably we see it expressed in permanent form in the obelisks of Egyptian solar temples, though these no doubt were retained merely as solar emblems ages after their use as actual instruments of observation had ceased. An upright stick, carefully plumbed, standing on some level surface, may therefore well mark the first advance upon the natural horizon. A knob at the top of the stick will be found to render the shadow more easily observed.

The careful study of this instrument will enable the meridian line to be marked with some considerable exactness. This should be done by taking an observation at some time in the morning, some hours before noon, drawing a circle with the base of the stick as centre, and the length of the shadow as radius, and then in the afternoon watching till the tip of the shadow again lengthens itself to exactly reach the circle. We shall find the north point lie midway between the two positions of the shadow. Here again we must trust not one observation but many, and the mean will give us a very close approximation to the true meridian.

The date of the summer or of the winter solstice would not be very readily ascertained from such an instrument —the very word solstice intimating that the change in the sun's position at that season is scarcely perceptible. But the time of the equinoxes can be fixed with sufficient exactness, since the length of the noonday shadow of a six-foot rod will vary in our latitude more than an inch a day at that time of the year.

A far exacter instrument for the observation of the sun can be made with the very slightest trouble; a light tube, 5 feet 4 inches long, made either of tin or of pasteboard, and covered at one end with a cardboard disc, with a pinhole one-sixteenth of an inch in diameter, carefully perforated in its centre, and at the other with a cap of oiled paper, will enable the sun to be observed with great ease. If this tube is directed to the sun an image of the sun will be formed by the pinhole on the oiled paper some six-tenths of an inch in diameter, and if a cardboard disc some ten or twelve inches in diameter is fixed to the tube—the tube passing through the centre of it—so as to screen the observer from the rays of the sun, he will find the sun's image on the oiled paper quite bright enough to observe, and much better defined than the shadow given him by the rod.

The next step would be to fit the tube with a graduated circle. The material of which the circle should be made and the manner in which it should be graduated may be left to the ingenuity of the student. Protractors of horn, metal, glass, or card can be very easily purchased and may well serve the purpose. The reading of the circle may be accomplished in one or two ways ; the circle may be fixed firmly to the telescope so as to turn with it, and the altitude of the tube may then be read by a plumb-line dropped from the centre of the circle across its circumference ; or the circle may itself be fixed in one position with respect to the vertical, and the tube may be turned round upon the same centre as that of the circle. In this case the tube should be supplied with pointers to read on the circle.

The tube being provided with a vertical circle and constructed so as to turn in a vertical plane, should also have its stand so arranged that it may turn in a horizontal plane also, and it should be fitted with a second circle, the centre of which is the pivot on which it turns. This circle must be fixed in the horizontal plane, and our instrument will then be a rough model of an altazimuth.

Its first use will be to determine the meridian—by taking an observation in the morning, reading both circles—then in the afternoon, waiting until the sun had descended to the same altitude a second time, and then reading the azimuth circle again. To set the telescope to' the azimuth midway between these two azimuths would be to set it roughly in the meridian. Here again the observations should be repeated many times, and the mean should be taken as the true south point.

The south point once found, the observation of the varying altitude of the sun at noon from day to day throughout the year would be a simple and easy matter. At midsummer and midwinter the meridian altitude of the sun will not vary perceptibly for a fortnight or more, so that we shall obtain a number of observations for the greatest and least height of the sun. Half the difference between these two must plainly be the obliquity of the ecliptic; and the altitude which is the mean of these two extreme altitudes must be the altitude of the equator, that is to say of the sun when it is at the equinox. The date of the equinox will be determined to the nearest day without any difficulty, for if we set our tube in the meridian, and pointing to the equator—in other words at an altitude equal to the co-latitude of our station—a single day's variation in the height of the sun at the time of the equinox will make a change in the position of the sun in the field of our pinhole tube of about four-tenths of an inch—an amount which the very roughest of observers could not overlook.

Such an instrument, simple as it is, would therefore enable the observer to determine the date of the equinox to the nearest day, and consequently the length of the tropical year, and also the obliquity of the ecliptic and the co-latitude of the place of observation. The exactness with which these could be determined would depend upon the skill and patience of the observer, who could, ere long, if he were sufficiently exact, begin to detect causes of irregularity in his results, some due to defects in his instrument, and some due to causes apart from that apparent motion of the sun which it was his first object to determine. The detection of these, and the discovery of their cause, will give a keen delight to anyone with a true observer's spirit, especially when he finds that a proper allowance for them brings his observations into ever closer and closer accord.

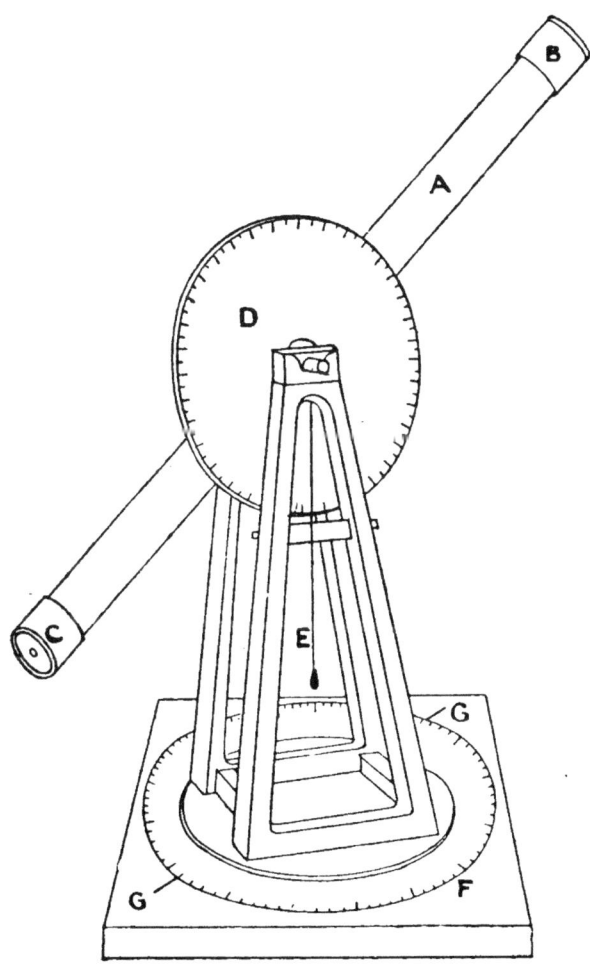

FIG. 24.—Pinhole Altazimuth for Observation of the Sun.

A, Pinhole tube; B, cap for pinhole; C, cap with screen for image; D, altitude circle; E, plumb line; F, azimuth circle; G, G, pointers for reading azimuth circle.

CHAPTER II.

Morning and Evening Stars.

It is one of the necessary penalties of the modern tendency towards city life that we are dissociated more and more from that close intercourse with nature which was open to our forefathers. How few of us ever care to watch that great spectacle which was to them so full of wonder and of awe, the silent, ceaseless procession of the starry heavens! It was not merely the sight of the thousand flashing gems of the midnight sky that impressed them, or their differences of colour and lustre, or the weird manner in which they were distributed; there was something more striking than all this, and that was their ceaseless movement. There was the wonder; that motion was so stately, so regular, so unceasing. "Without rest, without haste" they moved; no star ever left its appointed place in the celestial host, nor ever strayed from its appointed path. It was a nightly miracle, a miracle both of order and power. The thoughts to which it gave rise lay at the root of many an ancient myth, and inspired many a poetic outburst, chief amongst which stands the grand 19th Psalm. Nor, though the secret of that regular motion is understood to-day, is the sight of the movement of the vast cosmical machine less impressive even now to any mind that can rise to some slight realisation of its true meaning.

But the observers of those early ages had other thoughts besides those of awe and wonder as they contemplated the nightly march of the heavenly host; there came a time when they had familiarized themselves with the different stellar groupings, and they saw that different constellations ruled the night watches at different seasons of the year.

After a long night watch in which the stately march of the heavenly host has been followed hour after hour, some stars moving round the pole in narrow circles, others passing out of view in the west, whilst new ones from the under-world are ever mounting upward in the east, there comes a time when a change begins to be perceptible in the latter quarter. Little by little the eastern sky begins to brighten and the stars to pale, and now perchance it may be that a star more magnificent than any that has shone the long night through, comes upward, queen of the night, as the night is coming to an end. As it rises higher, so the glow in the east becomes brighter, and the fainter stars die out, until at length all have disappeared in the increasing daylight and the sun itself arises, the bright herald which preceded it being the only one of all the starry company that remains visible till the sun's appearance.

The planet Venus is the only object that in this completest manner can act as the forerunner of the sun, and it rightly, at such seasons, has a special claim to the title of the " Morning Star." But it is only now and then that Venus is so placed as thus to act as the solar herald. At other times the *rôle* has to be filled by stars of a meaner rank. But it is well worth noting what stars are the last

L 2

to rise before the dawn drowns them with its growing light. The next morning it will be found that those same stars are visible just a little longer, and the next morning longer still, and so on, until some other star shows itself as the one to climb from the eastern horizon just before the opening daylight becomes strong enough to overcome its shining.

Such stars are each in their turn " morning stars," and their first appearance in the faint dawn-glow of the east before sunrise is the "heliacal rising" which was made much of by ancient astronomers and poets. Not without reason, for it is an observation of very considerable exactness, and one which requires absolutely no instrument ; not even the simple one of an obelisk or an upright spear, still less of the solid masonry of a "solstitial" or " equinoctial temple," or the huge trilithons of a Stonehenge. And it fixes the return of the sun to the same part of the heavens at the end of a year quite as exactly as those more cumbrous contrivances could do, if indeed they were ever used for such a purpose.

Corresponding to the "morning stars" are the "evening stars," the stars which are seen in the western twilight just before they set; the gathering darkness permitting them to be just glimpsed for a minute or two before they follow the departed sun down to the under-world. This constitutes their " heliacal setting," and its observation supplements that of the " heliacal rising."

All stars are not by any means equally suitable for observation of their heliacal risings and settings. It is of course clear to begin with that the circumpolar stars, the stars which never set, are quite useless in such a connection. Nor are the stars in the extreme south, which only rise at the best a few degrees above the horizon, and only remain visible a very few hours, well suited for the work. The best stars are those which are not situated at any very great distance from the equator, either north or south, but which lie between the tropics of Cancer and Capricorn. Rigel, Sirius, Procyon, Arcturus, Alpha Serpentis, Alpha Ophiuchi are fairly well placed for observation in England, both at rising and setting, but as a general rule we may take it that the time of the spring equinox is the best for observing heliacal settings, and that of the autumnal for observing heliacal risings. The reason of this will be seen at once from the accompanying diagrams, which represent the effect of the sun's motion in declination at both rising and setting, at the spring equinox; at the autumnal equinox the conditions are, of course, exactly reversed.

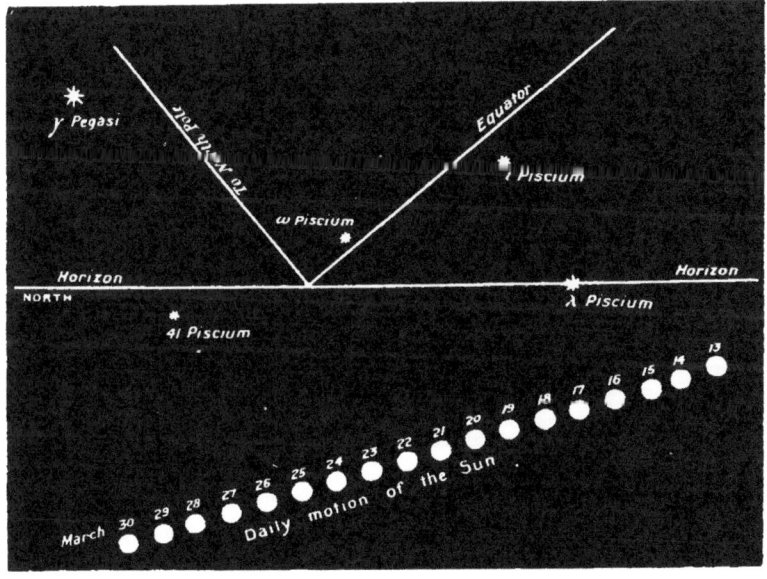

FIG. 25.—Positions of the Sun below the East horizon, near the Spring Equinox, at the rising of λ Piscium.

If the sun moved always in the equator, then a star, which rose on one day at precisely the same time as the sun, would rise about 4 minutes earlier than the sun on the next, and so on day by day'. But at the time of the spring equinox, the sun is moving northwards at the rate of about 24 minutes of declination every day. Consequently the sun rises about two minutes earlier every day, and the stars which rise at dawn only gain therefore upon the sun 4 m.—2 m.; that is to say, two minutes. If the

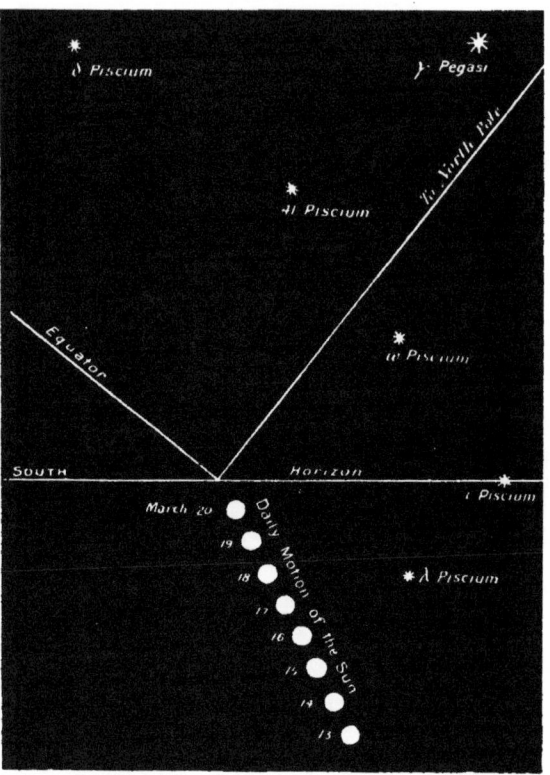

Fig. 26.—Positions of the Sun below the West horizon, near the Spring Équinox, at the setting of ι Piscium.

star is north of the equator, the sun is also coming nearer to it day by day, and hence that region of the sky becomes brighter and brighter at dawn. The result is that there may readily be a good deal of uncertainty about the determination of the day of the heliacal rising of such a star under these conditions. For the setting, the conditions are reversed. If the sun were always in the equator, a star which set with the sun one day would set 4 minutes before it the next, and so on day by day. But since the sun is moving northward at the spring equinox, it sets about two minutes later each day, and the star therefore would set before the sun 4 m. + 2 m., or six minutes the next night, and so on night after night. If the star was south of the equator, the sun would be moving away from it in declination, and under these circumstances we should get the sharpest determination of the date of heliacal setting. So the autumnal equinox gives the best opportunity for determining heliacal risings, specially for stars north of the equator, and the worst time for observing heliacal settings, especially for stars south of the equator.

It follows from these considerations that the Pleiades, in spite of the frequent reference to their heliacal rising in classical poetry, were not well placed for this observation, for during the 5000 years that astronomy has been in active operation as a science, they were never so far from the vernal equinox as at present. It must, however, be borne in mind that the nearer to the equator the place of observation, the more exactly could the observation of heliacal rising be made. For the nearer the equator is approached, the more nearly are the apparent paths of the sun and stars perpendicular to the horizon. They, therefore, seem to rise in the sky much more sharply, and dawn and twilight last for a much shorter space of time. A similar remark applies in its degree to Sirius, which, 5000 years ago, would have been by no means a very good star to observe at heliacal rising here in England. But the land where Sirius was made much of for this purpose was Egypt, the most southern of all the ancient civilisations, and in that marvellous climate the star may often be seen now to rise sheer from the waters of the Red Sea. Under such circumstances the return of the year could be fixed with all desired exactness by the rising of Sirius, even when it was on the vernal colure.

There is a great interest for those who care to try to enter into the thoughts and conditions of the past in repeating for oneself the very observations which were so all-important thousands of years ago. And there is a real value in repeating them to-day. If three or four observers were independently to observe the heliacal risings and settings of some two dozen zodiacal stars for three or four years, we should gain a great deal of needed information as to the actual amount of precision of which the method is capable, and some needed light would be thrown upon its practical efficiency. But more than this, the work would prove very effective in teaching the observer to pick up quickly minute points of light in a bright sky. This training would be found afterwards of great value in more ambitious fields of astronomy. The variable star observer would find it of use in enabling him to discount the effect of bright moonlight on his observations; the double star observer would find his eye quickened to detect a faint companion in the glare which a bright star gives in the field of a great refractor.

But more than this, the work affords a chance—a rare one it must be admitted, but a chance not to be despised —of making a discovery of a most interesting kind. Occasionally comets passing through our system so approach us as not to be seen until they are close to the sun, and these are usually comets of exceptional interest, both from the character of their orbits and from their physical behaviour. Such objects the persistent observer of heliacal risings and settings is—from the nature of his work, and from the special acuteness in detecting objects in the twilight sky which it will have given to him —the most likely to be the first to discover. The great comet of the autumn of 1882, for instance, was first seen at sundown, the great southern comet of 1901 was detected just before sunrise ; in both cases only a very short time before the comet passed into conjunction with the sun, and was temporarily lost to view. In both cases a regular " heliacal " observer would have stood a good chance of being two or three days ahead of anyone else in the discovery, to the gain of his own reputation and of astronomical science.

The three stars mentioned earlier, Sirius, Procyon and Rigel, may be observed to rise during the latter half of August ; their settings fall in April and May. They are thus emphatically winter stars, Rigel coming to opposition on December 11th, Sirius on December 31st, and Procyon on January 12th. In the case of these stars, the opposition falls nearly midway between the date of rising and that of setting. Not so with stars near the equinoctial colures. Thus Spica rises at the end of October, comes to opposition on April 12th, and sets in the middle of August. Hamal, on the other hand, rises near the end of May, comes to opposition on October 26th, and sets in the middle of April. The Pleiades, too, which Hesiod reports as being invisible for forty days, are now invisible here in England for over sixty, from the end of April to the beginning of July, but the date of conjunction does not fall in the middle of this period but on May 19th.

A very interesting side question would probably arise in the course of the systematic observation of heliacal phenomena, and that is the influence of colour on the visibility of stars of a given brightness in a bright sky. Orion and the neighbourhood are particularly rich in stars suitable for such a comparison. The orange tint of Betelgeux, the golden colour of Aldebaran, the slight suspicion of green about Bellatrix, the steel-blue of Alnath and of the giant Sirius, the white of Procyon, and the contrast which is so obvious between the tints of Castor and Pollux, afford ample materials for a very delicate and interesting research, and one which it still remains to make.

STARS FOR HELIACAL OBSERVATION.

AT SETTING.	AT RISING.
February—ε Pegasi, *Enif.*	August—β Orionis, *Rigel.*
March—α Pegasi, *Markab.*	α Canis Minoris, *Procyon.*
γ Pegasi, *Alaenib.*	α Canis Majoris, *Sirius.*
α Piscium, *Okda.*	α Cancri.
April—α Arietis, *Hamal.*	September—α Leonis, *Regulus.*
β Orionis, *Rigel.*	β Leonis, *Denebola.*
η Tauri. *Alcyone.*	October—α Bootis, *Arcturus.*
May—α Canis Majoris, *Sirius.*	α Virginis, *Spica.*
α Tauri, *Aldebaran.*	November—α Serpentis, *Unuk.*
α Canis Minoris, *Procyon.*	α Ophiuchi, *Rasalague.*

CHAPTER III.

THE MARCH OF THE PLANETS.

THE nightly procession of the stars across the sky was not the only celestial movement which impressed the observers of old. The coming and going of the planets held an even stronger attraction for them, and the mystery of their wanderings, which seemed at first sight to be so lawless, was the subject of much deep speculation. But this field of work has been so completely occupied in modern times by the transit circle and allied instruments that it is now hopeless for the "Astronomer without a Telescope" to dream of obtaining results of any real value. The only thing that he can do is to imitate his fore-runners and to familiarise himself with the apparent motions of the planets as they present themselves to the naked eye.

In this work he will find that the five planets within the reach of his unaided sight divide themselves into three groups. The first group includes Mercury and Venus, which moving in orbits interior to that of the earth, can never come into opposition to the sun, but oscillate backwards and forwards on either side of him. Both of them are therefore most easily observed when they are at their elongations. . Of the two, Mercury, being much the less bright, since he is smaller, and has a less reflective surface than Venus, is by far the more difficult object, a difficulty increased by the fact that his elongation cannot under the most favourable conditions amount to 28°, whilst the greatest elongation for Venus is nearly 48°. For reasons corresponding to those which were considered in the preceding chapter as regulating the most favourable conditions for observing heliacal risings and settings of stars, the most favourable position for Mercury to be seen as an evening star is when his eastern elongation occurs near the spring equinox; his most favourable position for observation as a morning star is when his western elongation occurs near the autumnal equinox.

Comparatively few dwellers in England have seen Mercury with the naked eye, unless they have persistently searched for him, and hence the idea has arisen that there was something very remarkable about the know-ledge which the ancients had of this planet. But as Colonel Markwick remarked in a paper in KNOWLEDGE (July, 1895, p. 152)—"To anyone who has seen that planet in the latitude of Greece, it would seem a most remarkable thing if the planet had not been seen and noted as such." This proceeds partly from the climates of Chaldea, Egypt and Greece being far better than that which we possess here in England. But in addition to this the greater nearness to the equator of these countries involves that the daily path of the sun and planets is more nearly perpendicular to the horizon, and the twilight much shorter than with us, giving an immense advantage over us for observing this planet. The accompanying little diagram, reproduced, by Colone Markwick's permission, from the paper just cited, shows the positions of the equator and of the tropics of Cancer and Capricorn, which mark roughly the limits between which the paths of the two planets are confined, as seen at sunset in the latitudes of 52° N. and 36° N. respectively.

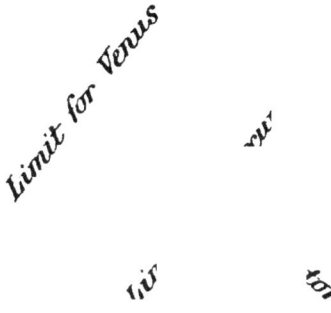

*Section of Equatorial Belt
looking due W. at Sunset   Lat 52° N.*

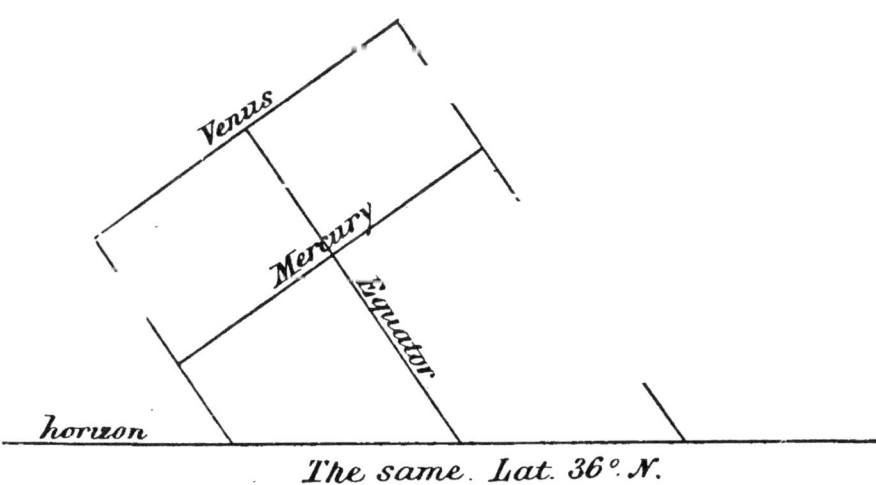

The same. Lat. 36° N.
1 in - 30°

FIG. 27.—Positions of the Equator and Tropics, relative to the West horizon, for Latitudes 52° and 36° N.

The interval between an east elongation of Mercury and the following west elongation is a little over six weeks, Mercury passing between the sun and the earth in the meantime. The interval from west elongation to east elongation, Mercury passing behind the sun, is a little over ten weeks, the mean periods being $43\frac{1}{2}$ days and $72\frac{1}{2}$ days respectively. The mean interval therefore from one elongation to the same elongation again is 116 days. Since therefore Mercury goes through the complete cycle of his positions with regard to the sun in a little under four months, there must be at least one favourable configuration in every year. In general there will be two or three. The attentive watcher therefore should have no difficulty in catching sight of the elusive little planet at one elongation at least in each year.

The motions of Venus are similar in character to those of Mercury, but are performed more slowly, and the arc through which she swings is a wider one. The mean period from east elongation as an evening star to west elongation as a morning star is 143 days, whilst it is 441 days from west elongation round to east; the entire synodic period therefore being 584 days. There is of course no difficulty at all in recognising Venus when she is near her elongations, for she is then the most brilliant star in the sky, but some interest attaches to her changes of brightness. These changes depend upon two circumstances: the one, the distance which she is from the earth, and consequently the apparent size of her disc; the other, upon the amount of that disc which is lighted up by the sun, in other words upon her phase angle. The disc is greatest when Venus is between the earth and the sun, but at such a time the face she turns to us is necessarily in darkness. On the other hand, when she is in superior conjunction, that is to say on the further side of the sun, her disc is entirely illuminated, but it is then at its smallest size. In both cases she is invisible since she is in full sunlight.

At elongation her disc is much larger than at superior conjunction, so much larger indeed that though it then appears like a "half-moon," it gives us three times the light which it would do could we observe it when it was full. But as she passes from east elongation towards inferior conjunction, the increase in apparent size compensates, and more than compensates, for the decrease in phase, until half-way between the two positions her light is four times that at superior conjunction. From this point the effect of phase is greater than that of increase of size. The time of greatest brilliancy therefore is some 36 days before inferior conjunction, and again 36 days after. Then a period of 512 days will ensue before Venus returns a second time to her greatest brilliancy as an evening star. And as five times 584 days—the synodic period of Venus—is almost exactly eight terrestrial years, it follows that at intervals of eight years the times of greatest brilliancy recur on almost the same dates.

At her greatest brightness, Venus is so brilliant that there is no real difficulty in seeing her with the naked eye in full sunshine, or indeed at high noon. The hindrance is not the want of brightness of the planet, but the difficulty in picking up so minute a point of light as she appears to be, unless some means is provided for guiding the eye to the precise spot. When so found the first impression is—"How could I possibly have overlooked so bright an object?" The next impression, if the eye be turned from the planet but for a moment, is—"What a hopeless task it is to try and find it again!"

The statement has sometimes been made that the phase of Venus may be seen under exceptional conditions with the naked eye. Frankly I think this observation lies outside the limit of possibility; for Venus at her greatest brilliance is only about 40 seconds of arc in diameter.

Now 40 seconds is practically the limit for distinct defined vision, and it is very improbable that the precise shape of Venus could be detected under such circumstances. It would, however, be interesting to know if any one could detect that the planet was not round. It is just possible that an impression of elongation might be given, though this would show exceptionally keen sight.

The second group of the planets includes Saturn and Jupiter, which both move in orbits far outside that of the earth. From the point of view of the naked-eye astronomer, neither is of very great interest. Their movements do not greatly differ from those of the stars, amongst which they move but slowly. Like the stars they will have their heliacal risings, being seen as morning stars in the east just before sunrise. Rising earlier and earlier the time comes when they are visible for the entire night. But before they are in opposition to the sun, that is to say, are on the meridian at midnight, there is a striking change in their apparent motion amongst the stars. For the greater part of the year, Saturn is moving eastwards amongst the stars at an average rate of a degree in about eight days. Jupiter traverses the same distance in about half that time; but gradually as the time of opposition draws on, the speed of both planets diminishes, until Saturn comes to a stop about 70 days before opposition, and Jupiter about 60. Then for 143 days Saturn moves westward, and Jupiter for 122, both planets becoming stationary again at the end of the period, and then resuming once more their eastward march. This period of westerly movement or retrogression marks the time when the planet is nearest to the earth and therefore brightest, and it is at the middle of this period that the planet is in opposition, that is to say, is on the meridian at midnight.

Just as it has been asserted that the crescent of Venus has been seen with the naked eye, so it is also asserted, but on somewhat better authority, that the satellites of Jupiter have been seen at their elongations from their primary, and though of course it is utterly beyond the unaided sight to perceive the ring of Saturn, the claim has been made that Saturn has been observed as an elongated, not as a circular point of light. By a curious coincidence it happens that the diameter of Jupiter at opposition, the major axis of Saturn's ring, and the diameter of Venus at greatest brilliancy, are all very nearly equal, and all are very near the limit of defined vision. It follows therefore that the Saturn observation is as difficult as that of Venus, but the satellites of Jupiter are not quite so hopeless. The first and second are always too close to their primary to be seen apart from it, and they are probably too faint as well. But the third would be very readily visible if it were a solitary star, and at its greatest elongation from the planet it is distant from it $5\frac{3}{4}$ minutes of arc, one-sixth of the apparent diameter of the moon. Many people can separate ϵ_1 and ϵ_2 Lyræ which are considerably nearer to each other. The fourth satellite attains a distance at greatest elongation of nearly ten minutes of arc, a distance amply sufficient to separate it from Jupiter, but is by no means so bright an object as the third. The story of Wrangel, the celebrated Russian traveller, quoted by Mr. G. F. Chambers from Arago, that when in Siberia he once met a hunter who said, pointing to Jupiter, "I have just seen that large star swallow a small one and vomit it shortly afterwards," is a beautiful specimen of the traveller's tale. Wrangel explains the hunter's remark as referring to an immersion and subsequent emersion of the third satellite. It escaped Arago's notice that it takes the third satellite over a week to pass from one elongation to another, and that, further, as the satellite would have reappeared on the opposite side of Jupiter from that on which it disappeared, the hunter would have scarcely described the incident as he did.

An opera-glass of course easily shows the satellites of Jupiter, and one optically perfect should suffice to elongate Saturn when the ring is fully open or to show the phase of Venus at greatest brilliancy.

The movements of Mars are sufficiently different from those of the other four planets for him to be considered by himself; the chief features in his case being the length of time in which he remains out of sight on the far side of

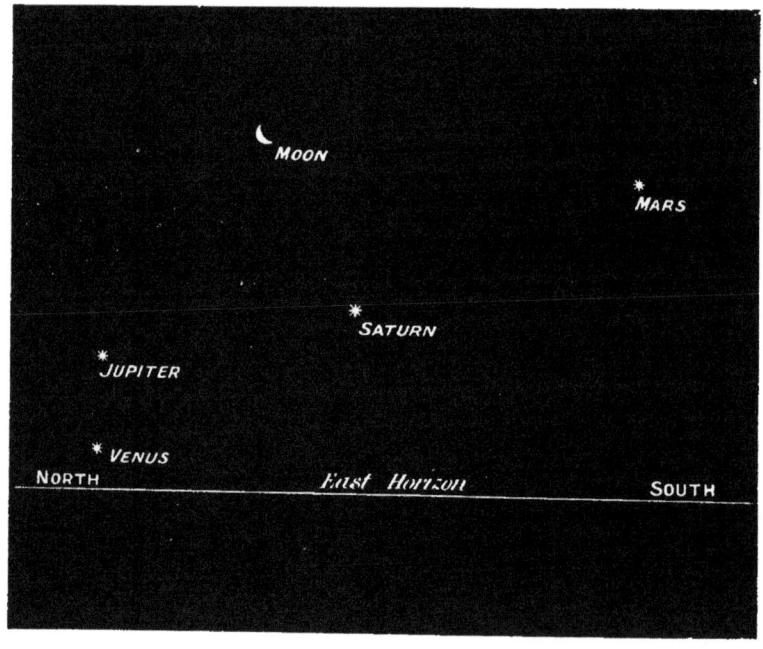

FIG. 28.—The East Horizon at 2 a.m. on the morning of 1881, June 22.

the sun, and the very great difference between his apparent size when nearest to the earth and when farthest from it. The synodic periods of Mercury and Venus stated above are 116 and 584 days respectively; those of Jupiter and Saturn are 34 and 13 days longer than an entire year, but the synodic period of Mars is 50 days more than two years. During the greater part of this time he is moving eastward amongst the stars at a pace of only a little less than a degree a day. As he approaches opposition his pace slackens down until, like Jupiter and Saturn, he becomes stationary and then recedes for a time westward. This period of recession begins about six weeks before opposition and lasts about six weeks after.

The chief interest of Mars, as a subject of observation without the telescope, is the opportunity which he gives, especially when he is in the more distant parts of his orbit, and therefore relatively fainter, to compare him as to his brightness with the stars near to which he passes. This is a work which needs doing, and which anyone could undertake, yet it has been almost entirely neglected.

Three other planets are just within the limits of visibility under favourable circumstances, Uranus, Ceres and Vesta. To these we may now possibly add a fourth, Eros. But while the three above mentioned might be seen at any opposition, Eros, if ever bright enough, will only be so at one opposition in thirty-seven years.

One feature of the march of the planets will always attract attention and give pleasure to the observer, even though no practical result be drawn from it; that is the way in which from time to time two or more of them will come into close proximity to each other, or it may be to some bright fixed star or to the moon. Thus on September 15, 1186, all the five chief planets were in conjunction together in the constellation Virgo to the east of Spica. In much more recent times, four of the five chief planets came together in the constellation Aries, Mercury being the only missing member, and this conjunction was made the more impressive by the presence in the midst of the group of the waning moon. This beautiful spectacle was witnessed in the early morning of June 22, 1881.

CHAPTER IV.

Sunspots and Moonspots.

THERE are only two of the heavenly bodies which present a disc to our ordinary sight, and the surfaces of which therefore we can study without a telescope. These are the sun and moon. It is, of course, absolutely impossible that observations thus made can in any way compete with those made even with a hand-telescope; but from the point of view of astronomical drill, as distinguished from actual research, there is something to be said for systematic work upon both of them. Jupiter as seen with a magnifying power of 50, Mars at a mean opposition with one of 100, Saturn with a similar magnification, present about the same apparent disc as the sun and moon do to the naked eye. There is therefore a real interest in seeing how much detail the eye can actually detect upon these two bodies. The limit of magnification possible for the efficient study of the surfaces of the planets is soon reached, and when an astronomical artist has done his very best with Jupiter, Mars, Saturn, or Venus, it would be an invaluable check upon his work if he would draw the sun or moon with a little instrument, and such small magnification as would give to its disc the same apparent diameter as had been presented to him by the planet which he had just been studying. The "Bulletin de la Société Astronomique de France" for 1900 contains a large number of such drawings of the moon, made with

NORTH.

EAST.

WEST.

SOUTH.

FIG. 29.—Photograph of Moon, taken 1902, March 22, 10h. 37m. 24s. G.M.T.

the naked eye, and the study of them is, I think, most instructive on a number of points which have been in dispute from time to time as to the condition of the surface of Mars.

The defining power of the eye is, of course, limited, and when a number of details are presented to it, each one of which is much too small to be defined separately, all together produce a composite effect to which each detail has contributed in its own degree. Now there certainly is a wide difference in the manner both in which such composite effects will impress different persons, and in the way in which they will record them in a drawing. And the study of the "personality" of astronomical artists should be a necessary precedent of the comparison and collation of their drawings. The drawings of the moon, given in the volume referred to above, are as widely different as any set of drawings that were ever made of Jupiter or Mars. Yet a careful comparison of them with maps or photographs of the moon will show that the forms given are not imaginary, but have a real relation to the lunar markings, whether they be skilfully represented or no.

The following drawings are from the volume cited, pages 277 and 505, and are by M. Maurice Petit and M. E. M. Antoniadi respectively. The latter comments on his observations as follows :—

"It is a work of immense difficulty to draw correctly all the "details that the naked eye reveals to us on the surface of our "satellite. It is above all things necessary that the moon should "be seen with the greatest possible distinctness. If the eye is not "emmetropic it will be necessary to select glasses of the proper "focus bringing the focus *exactly* on the retina. The accom- "panying sketch which is only a rough approximation was obtained "with concave glasses, allowing at least ten stars to be seen in the "group of the Pleiades ; it is the result of studies covering several "lunations. . . . . The darkest spots seemed to me (1), the Mare "Tranquillitatis, and (2) Mare Nubium. The Mare Serenitatis and "Fecunditatis come next. Mare Imbrium and Oceanus Procel- "larum are still paler. The little Maria Crisium,Vaporum, Humorum "and Nectaris, are reduced in size by irradiation, and present to the "naked eye an appearance corresponding to that of the Lacus Solis "and Lacus Lunæ of Mars as seen in a telescope. The white spots "of Copernicus and Kepler are very well seen, Aristarchus with "more difficulty, whilst Tycho, with its brilliant surroundings,

" occupies an immense white surface ; but . . . . the ring itself
" is not seen. Gueriké, Bonpland, Parry and Fra Mauro make up
" a bright island in the Mare Nubium. Lastly, the Mare Serenitatis
" is whiter in the centre by reason of the great white streak coming
" from Tycho which crosses it."

With regard to the sun, it may be thought that absolutely
nothing can be done without optical assistance.   But any

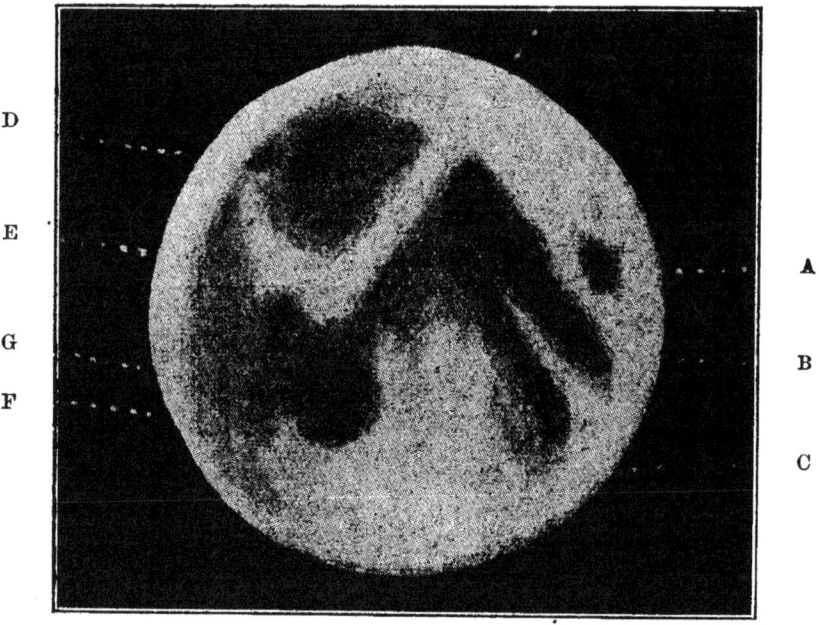

A, Mare Crisium.   B, Mare Foecunditatis.   C, Maria Nectaris et Tranquillitatis. D, Mare Imbrium. E, Oceanus Procellarum. F, Mare Humorum.
G, Mare Nubium. H, Apennines and Copernicus.

FIG. 30.—Drawing of the Moon by M. Maurice Petit.

observer who will give himself day by day to its patient
scrutiny will be astonished ere long at the result.   Of
course, in this case, the term " naked eye " is no longer
applicable, since the eye must be shielded from the overpowering light of the sun by a dark glass or the
sun's light may be admitted through a small hole into a

darkened chamber and received upon a white screen. But thus observed the sun at maximum activity will show a spot large enough to be *easily* seen on fully one day in four; not infrequently two or three spots may be seen at the same time. Had it occurred to the classical and mediæval astronomers to watch the sun systematically in this way, they would not only have detected the existence

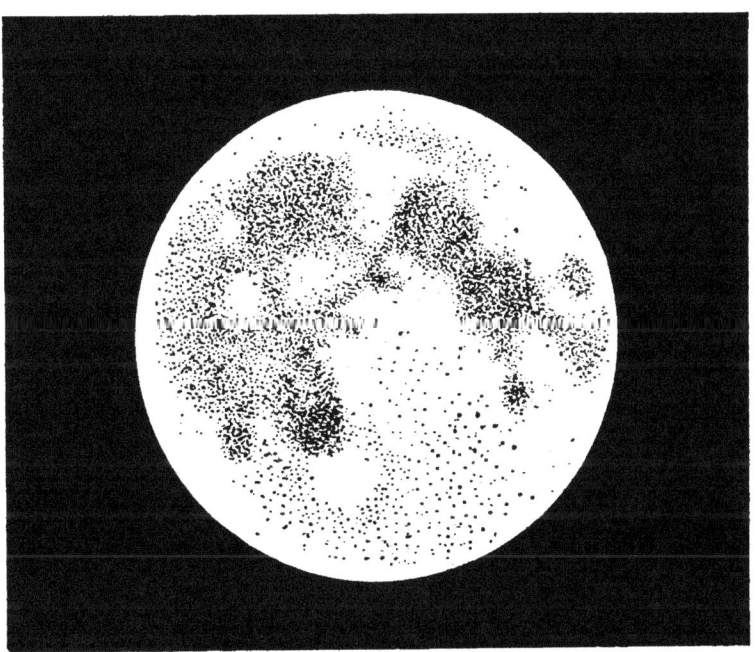

FIG. 31.—Drawing of the Moon by M. Antoniadi.

of spots on its disc, but have demonstrated as certainly as we know it to-day the period of the sunspot cycle, and the value to us of that information would have been incalculable. More than that, it would have been possible for them, from a long series of observations, to have fixed the solar rotation period fairly exactly and to have made a first approximation to the determination of the position of the axis.

The accompanying little table and diagram give a comparison between the number of days in each year during the last two decades in which there were spot groups the

Year.	No. of Days on which Spot-groups exceeding one-thousandth of the Sun's disc were visible.	Mean daily spotted area expressed in millionths of the Sun's visible hemisphere.
1882	78	1002
1883	129	1155
1884	91	1079
1885	68	811
1886	35	381
1887	17	179
1888	3	89
1889	3	78
1890	8	99
1891	32	569
1892	99	1214
1893	135	1464
1894	112	1282
1895	82	974
1896	58	543
1897	52	514
1898	42	375
1899	9	111
1900	1	75
1901	5	29

area of which as seen covered more than one-thousandth of the sun's apparent disc, and the mean daily spotted area expressed in millionths of the visible hemisphere as deduced from the photographic record at Greenwich. It will be seen that the first record comes as sharply to a maximum as the second, and falls off again as unmistakably to a minimum. Even some minor details of the second curve are faithfully indicated in the first. The

limit chosen—one-thousandth of the visible disc, or almost
exactly the apparent size of Venus in transit—would not
in all cases correspond precisely to the limit for the visibility of spots; but general experience has shown that the
limit selected corresponds very nearly to the limit of
visibility without optical assistance, differing from it in

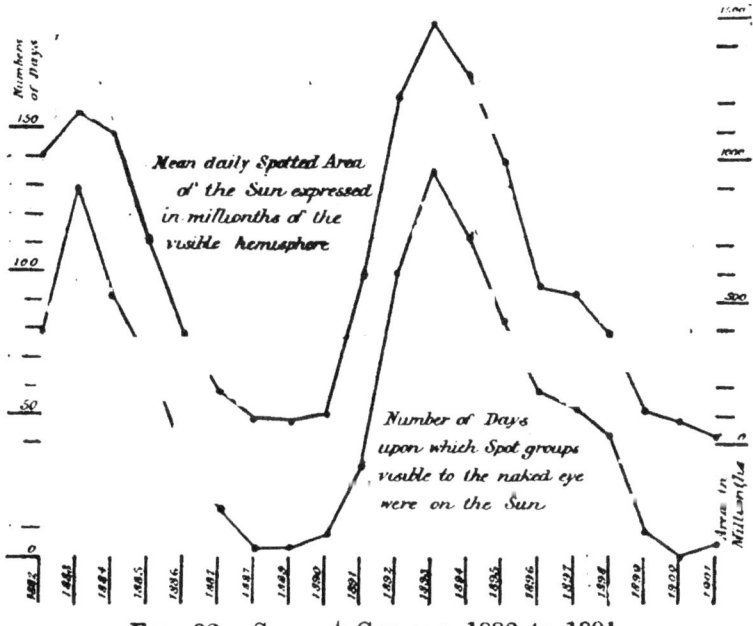

Fig. 32.—Sunspot Curves; 1882 to 1891.

under-stating, not over-stating, the number of groups
which could be seen.

Of course, there is no possibility now for the "Astronomer without a Telescope" to improve on the information
which the telescope can give us as to solar physics, but the
lesson which the foregoing table has to teach is a most
important one. Observations, as difficult and as apparently
hopeless as observations of the solar surface would have
seemed to be in the Middle Ages, may, if carried out
patiently and systematically, bear as rich fruit as solar
observations could have done even then. Even after the

invention of the telescope, it was not the optical power of his instrument, but the perseverance, with which he worked at a single object which revealed to Schwabe the secret of the solar period. He had no dream of the discovery before him when he set out upon his researches. His own expression was that he "set out like Saul, looking for his father's asses, and found a kingdom." There are kingdoms yet to be won, even in those fields of astronomy which the telescope cannot touch. In particular, changes which are periodic in character will reveal the fact and circumstances of that periodicity to observations carried on patiently and continuously, even when the amount of such changes at their maximum only just come within the utmost limits of the power of the instruments used in the work.

CHAPTER V.

A Modern Tycho.

In the Chapter on the "March of the Planets," I remarked, "This field of work has been so completely occupied in modern times by the transit circle and allied instruments that it is now hopeless for the astronomer without a telescope to dream of obtaining results of any value." Yet though this is the case, it has happened even in our own day that circumstances have precluded an astronomer, not merely from using optical means and modern instruments of precision, but even from availing himself of the results which others have secured by their means. This condition prevails in India where religious sentiment places an embargo upon the results of western science. The very essence of Hindu life is the faithful carrying out of a routine of religious observances, inseparably connected with the knowledge of the positions of the planets and stars. For this purpose a full and correct calendar is needed for the right ordering of every Hindu family. To Europeans astronomy is the pursuit of the very few, and the multitude know nothing and care less about the planetary movements. Such ignorance to the Hindu

would indicate a blameworthy indifference to the duties
of religion and to the requirements of society.

But a serious difficulty has been growing for many
years. Two hundred years ago Rajah Jey Singh noted
that the calculation of the places of the stars as obtained
from the tables in common use, gave widely different
results from those obtained by observation; the times
and magnitudes of the eclipses of the sun and moon, the
times of the risings and settings of the planets, and of the
appearance of the new moon, were all faulty; the ancient
Siddhantas having been left without revision from the
time of Bhaskara, the author of the Siddhanta-Siromani,
nearly six hundred years before. It was in consequence
of this discrepancy, keenly felt even then, that Rajah
Jey Singh obtained the permission of Mahommed Shah
to erect observatories in order to obtain new data by
which the tables should be corrected. Three of the five
observatories erected by Jey Singh are well known, those
of Jeypore, Delhi, and Benares. Of these, the one at
Delhi, established about the year 1710, appears to have
been the first.

The Gentur Muntur, or Royal Observatory of Delhi,
though barely two centuries old, is of a thoroughly
ancient type in its conception, and was intended for
naked-eye work alone. Just as at the Royal Observatory
at Greenwich the transit circle and altazimuth are considered the two fundamental instruments, so here at
Delhi the chief structures were evidently designed for
corresponding purposes.

The first object to catch the eye is the great gnomon,
the vertical section of which is a right-angled triangle,
with an hypotenuse of 118 feet, a base of 104 feet, and
a perpendicular height of 57 approximately. The face of
the gnomon therefore is parallel to the axis of the earth,

its angle corresponding to the latitude of Delhi. Up the middle of the gnomon runs a staircase, and right and left of the gnomon are great sectors on which its shadow falls ; these also being provided with steps over which the shadow took four minutes of time, corresponding to one degree of arc, to pass. A smaller structure near has some variations in its design, the gnomon being in the centre and flanked on either side by semi-circles sloping downwards from it towards the horizon. Further to the south of these gnomons, the chief use of which must simply have

Fɪɢ. 33.—The Gentur Muntur, or Royal Observatory of Delhi.

been to give the solar time, are two large buildings, which are evidently intended to serve as altazimuths. The two buildings are exactly alike in design and size, and they give the appearance from outside of being miniatures of the Colosseum at Rome. Within they are seen to be perfectly circular enclosures. The wall of each is pierced by three tiers of windows, thirty in each tier, the breadth of each window opening being precisely equal to the width of the wall between each pair of windows. The difference between the two buildings is simply a question of position, the two being so arranged that the windows of one command precisely those azimuths which are hidden by the wall of the other. In the centre of each enclosure is a pillar rising to the height of the enclosing wall, whilst from the circular wall thirty stone sectors are directed towards the central pillar, but do not quite reach it. The building is $172\frac{1}{2}$ feet in circumference, or 55 feet in diameter, and the sectors running from the wall towards the pillar are $24\frac{1}{2}$ feet in length. The windows, no doubt, were intended to enable the altitude and azimuth of any celestial object to be read off at sight, and for rough positions no doubt it did fairly well.

But the chief interest of Jey Singh's observatories lies for us in the fact that they recall a time when astronomers sought for exactness by the erection of huge structures of stone. Of these the Great Pyramid is by far the greatest and most perfect example. The north shaft, pointing to Alpha Draconis, the pole star of the period, the grand gallery which may very well, as Proctor suggests, have been used as a vast transit chamber, give evidence of its astronomical purpose, whilst the care with which the exterior of the Pyramid has been oriented, would fix most delicately the dates of recurrence of the spring and autumnal equinoxes. This would be seen from the circumstance that on one day in spring the north face of the Pyramid would be in shadow at sunrise, and the next day the south face, whilst the reverse effect would be noted in the autumn. The exact days of the equinoxes would be thus pointed out, and the return of the year to the same point would be marked with great precision.

Britain has its own monument—Stonehenge—which has been claimed as, if not indeed an astronomical observatory, at least an astronomical temple, and many attempts have been made to determine the date at which it was erected. The difficulty, not to say the impossibility, of solving this problem in the present state of the monument may be inferred from the fact that the dates which different observers have deduced for its erection extend over a period of more than 2000 years.

The best work of astronomy, even in the pre-telescopic age, was never done in edifices like these. Nor, indeed, does it require much knowledge of human nature, essentially the same 5000 years ago as to-day, to see that the true secret of the Pyramid, the amply sufficient cause for its building, was the vanity of the ruling Pharaoh. We get a graphic hint of this in the narrative in Genesis of the founding of a yet earlier observatory, for so no doubt it was. "They said, Go to, let us build us a city and a tower, whose top may reach unto heaven, and let us make us a name." Alike at Delhi, at Ghizeh, and on Salisbury Plain, as by the Euphrates, to "make a name" was, beyond doubt, the exciting motive. Astronomers may have been employed to superintend the work, astronomy, or the cult of the celestial bodies may have been the excuse, but the real motive was ostentation.

But the work for which the pretentious buildings of the Rajah of Amber were designed, has been more efficiently accomplished quite recently by very humble means, and by a recluse in an obscure village.* Chandrasekhara Simha Samanta is a near relative of the Raja of Khandapara, one of the tributary chiefs of Orissa. At the early age of ten, having been taught a little astrology by one of his uncles, he became most anxious to measure on his own account the positions of the stars in their

---

* *See* KNOWLEDGE, Vol. XXII., p. 257. Siddhanta-Darpana.

nightly movements, and by the time that he was fifteen years of age and had learned to calculate the ephemerides of the planets and of the risings and settings of stars, he was deeply disappointed to find how great was the discordance between his calculations and what he actually observed. It was no wonder that he found discordances; no two of the current Hindu almanacs agree in their predictions, and one of the most widely circulated of the Bengali almanacs may be as much as 4° out in the longitude of a planet.

In this difficulty Chandrasekhara had to work out his problem unaided. He had to make his instruments for himself, to some extent he had to devise them. The one of which he was fondest is a tangent staff consisting of a thin rod of wood twenty-four digits long, at the end of which is fixed another rod at right angles in the form of T. The cross piece is notched and also pierced with holes equal to the tangents of the angles formed at the free extremity of the other rod.

For many years he has been carefully revising the Siddhantas in order to bring them into conformity with his observations made at the present period, and he has been able to obtain a most astonishing degree of accuracy in his results. Thus, the sidereal period for Mercury is only 0·0007 days different from that adopted by European astronomers; for Venus it is only 0·0028 days. The mean inclinations of the orbits of the planets to the ecliptic are correct to about a minute of arc. The errors of the ephemerides computed from his new constants are reduced to about one-tenth of those in some of the most widely circulated Hindu almanacs. In his discussion of the moon's motion, he made the discovery—independent and original on his part—of the lunar evection, variation and annual equation, which found no place in the earlier

Siddhantas. In much of his work he had the advantage of comparing his observations with those of Bhaskara, made more than seven hundred years earlier; not indeed that the latter had recorded his actual observations, but it was possible to ascertain what they must have been from the planetary elements which he had deduced from them. Nevertheless, to have obtained such important results and so high a degree of accuracy, by naked-eye observations and with entirely home-made instruments, and in the utter absence of modern book learning, is a striking illustration of what resolution can effect.

Chandrasekhara has been compared to Tycho Brahe, and the comparison is in many ways a just one, though the recluse of Orissa lacked many of the advantages possessed by the noble Dane. As to the accuracy of Tycho's work, it will be remembered that Kepler was led to the first of his three great laws by finding that his theory of the circular motion of the planets differed from an observation of Mars by Tycho by eight minutes of arc—but one-fourth of the moon's diameter. Kepler concluding that it was impossible that so good an observer could be in error to this extent, abandoned his hypothesis and tried that of motion in an ellipse.

In the recluse of the Orissa village, we seem to see reincarnated, as it were, one of the early fathers of the science, long centuries ere the telescope was dreamed of, as he grappled with the problems which the planetary movements offered to him for solution. More than that, he affords an example of the achievements within the reach of the naked-eye astronomer, and a telling illustration of the precision which patience and practice can give to hand and eye. And these are always needed. For be the telescope ever so good and powerful, still that which is by far the most important is the man at the eye-end.

SECTION III.

ASTRONOMICAL OBSERVATIONS WITHOUT A TELESCOPE.

ASTRONOMICAL OBSERVATIONS WITHOUT A TELESCOPE.

CHAPTER I.

METEORS.

IT is an old saying, of the truth of which we are often reminded by our daily experience, that what is everybody's business is nobody's business. Work which someone is obliged to do, or is paid to do, gets done. Work too which is only open to a few to undertake also generally finds that some of that few will undertake it. But that which is open to everybody and yet to which no one is appointed, no one is driven, hangs fire and is left undone.

To take one example, one of the very earliest achievements of astronomy was to determine the length of the year. This was done long ages ago, earlier than we have any record. But it was a necessary, or at any rate a very practical and useful work, and consequently was done at an early epoch. Take again a modern instance —the observation of double stars. This is a work which is by no means within everybody's reach. A powerful telescope, well mounted, clock driven, and furnished with a good micrometer, is the luxury of the few. But in spite of, perhaps we should rather say because of this restriction, double star observation has always found a number of ardent followers. So that, although it is but 120 years since this branch of astronomy took its rise, it has already made a most amazing progress.

On the other hand, the various branches of naked eye astronomy, branches open to every one who has eyes to see and a good atmosphere, have been left almost unworked. The departments of meteoric and variable star astronomy are the only two in which great and substantial progress has been made, and in both cases such progress has been the work of the last few years. In meteoric astronomy that progress has been especially striking. It therefore justly comes first of all the subjects for study open to the astronomer who has no optical assistance at his command; the more so that no other objects can be so easily or frequently observed by him as meteors; none can afford him such an opportunity for really useful work.

It is very striking in looking back into astronomical records to note how very recent is most of our information concerning meteors. For thousands of years men have been aware that there were " wandering stars to whom was reserved the blackness of darkness for ever." At times, too, they would come, " not single spies but in battalions," in such numbers and with such brightness as to compel attention and create the deepest astonishment and fear. But for all those ages it does not seem to have occurred to anyone to try and *observe* them; that is to say, to record such facts about them as it was possible to ascertain during the brief moments that they shone.

There is an immense gulf between the mere admiration

of the phenomena of nature and their observation. The first is utterly unfruitful; long generations of men pass, each having seen the same kind of event, and yet the accumulated experience of ages leads to nothing. But, on the other hand, let one man, or, better, let three or four give a few years to the careful, steady record of everything that they can ascertain about some phenomenon, however unpromising, and what marvellous facts leap into light!

How utterly ignorant even recognised authorities were but sixty years ago may be seen from the following quotation from a standard text-book bearing the date 1840.

> "The Falling Stars, and other fiery meteors, which are frequently seen at a considerable height in the atmosphere, and which have received different names according to the variety of their figure and size, arise from the fermentation of the effluvia of acid and alkaline bodies, which float in the atmosphere. When the more subtile parts of the effluvia are burnt away, the viscous and earthy parts became too heavy for the air to support, and by their gravity fall to the earth.
>
> "On the 13th of November, in the year 1833, a shower of meteors fell between lon. 61° in the Atlantic Ocean, and lon. 100° in Central Mexico, and from the North American Lakes to the southern side of Jamaica. These fireballs were of enormous size; one appeared larger than the full moon at rising. They all seemed to emanate from the same point, and were not accompanied by any particular sound. It was not found that any substance reached the ground so as to leave a residuum from the meteors."

It did not seem to occur to the writer of the above description that the circumstance which he mentions—namely, that the meteors "all seemed to emanate from the same point," itself proved that the meteors were entering the atmosphere from outside and were moving along parallel lines at the time of their entry.

The great display referred to above, however, was the foundation of modern meteoric astronomy. So magnificent a spectacle as was then witnessed not only attracted thousands of gazers, it caught the attention of men who were resolved to use every possible opportunity for learning.

The enormous numbers of meteors seen in the November shower of 1833 rendered it manifest that on that occasion, at any rate, the falling stars seemed to have their origin in a single point of the heavens, and therefore it became an important point, whenever a meteor was seen, to note exactly the direction of its flight. Humboldt, who had himself seen the great November shower of 1799, writing in 1844, recognized four points in the heavens from which meteors seemed to fall, and drew attention, though with some hesitation, to the reasons for thinking that the November shower was only occasionally to be seen in great force. Sir John Herschel, about the same time, recognised only two showers, those of August (the Perseids) and those of November (the Leonids). Now by the labours of a very few observers, one of whom, Mr. Denning, may be said to have outweighed all others put together in the value and number of his results, we know of many hundreds of radiant points, whilst the researches of Adams and Schiaparelli have enabled us in some cases to trace the meteor streams in their path, not only far beyond the spread of our own atmosphere but to the very limits of the solar system, and they have been shown to be not mere distempers of the air, but bodies of a truly planetary nature, travelling round the sun in orbits as defined as that of the earth itself.

How has this great advance been made? Simply by

careful, patient, intelligent observation. First of all by carefully noting the points in the sky where the meteor was first seen and where it disappeared. This requires a thorough knowledge of the constellations, as indeed all naked eye astronomy does, and great quickness of observation. The meteor worker must be able to fix the two extremities of the path at a glance and to remember them faithfully until he can in some way or other note them down. Mr. Denning advocates the use of a perfectly straight wand, held in the hand, and projected upon the path of each meteor immediately it is seen. The direction and position of the path relative to the stars amongst which it lies can then be noted with considerable precision, and the path marked down upon a globe or chart.

For this he will require a certain amount of what we may term apparatus, either a celestial globe or a set of star charts. The choice of the latter is of importance, as no possible chart can show the entire sky without grave distortion some way or other; and more important for the present purpose, there is only one projection which will give a straight line on the chart for all great circles or parts of them; that is to say, for all lines which impress us as straight lines as we see them on the sky.

The observer's first duty, therefore, is to acquaint himself with the constellations; his next, by repeated and persistent effort, to learn quickness and correctness in fixing the extreme points of the meteor paths. This done, he will recognise that there are several other features in which meteors appear to differ, the one from the other. His observations of the paths will soon show him that the length of a meteor path varies greatly;

he cannot fail further to notice that the apparent speed with which it travels varies also. To the record of the path, therefore, should be added the determination of its length, which of course can be read off from the globe after the track has been marked down upon it, and the time which the meteor took to traverse it. And as for comparison with the records of other observers it is essential to know when the meteor was seen, this should also be noted as well as the duration.

The actual meteors themselves also have their individual characteristics. Some leave phosphorescent streaks behind them, others trains of sparks. More striking than anything else is the enormous difference in brightness, from one like the meteor alluded to above, "larger than the full moon at rising," down to others only just visible to the naked eye.

All these particulars should be duly and regularly recorded; the record taking the following or an equivalent form :—

1. Date, hour, and minute of appearance.
2. Brightness; in terms of stellar magnitude, or if very bright, in terms of the brightness of some planet visible at the time.
3. R.A. and Decl. of the point of first appearance of the meteor.
4. R.A. and Decl. of the point of disappearance.
5. Length of path.
6. Duration of visibility.
7. Chief characteristics, such as its colour, whether it showed streak or train, etc.
8. Radiant point; when a sufficient number of meteors have been observed for this to be determined.

Mr. Denning, in a paper on "Meteoric Fireballs,"

appearing in the *Quarterly Journal of the Astronomical Society of Wales* for November, 1898, makes the following amusing, but none the less earnest, appeal to be supplied with these necessary details instead of the flamboyant but useless descriptions so common in the ordinary accounts of fireballs or other bright meteors :—

"If amateurs generally paid attention to these details and really sought to secure that exactness which is so necessary in these observations, there is no doubt that this attractive, though difficult, field of practical astronomy would soon reap material benefit. In recent years there has scarcely been one out of ten fireball descriptions which has proved of utility. They have commonly lacked definite reference to the essential features. If a computer has only two full and accurate observations of a meteor, he can determine its real path in the atmosphere (including the particulars of its height, radiant, velocity, &c.) in considerably less than an hour. But when he has fifty or sixty imperfect and erroneous observations of a similar object he usually fails, after a laborious discussion extending over several days, to derive anything satisfactory from them. The discordances which such faulty observations present often lead him to conclude that in order to harmonise them several large meteors must have appeared at about the same time.

"I will quote two examples (omitting names) of published descriptions of meteors, one of which is of little utility, though couched in language well calculated to make our hearts throb with sentiment, and the other really valuable, though quite lacking in poetic fire :—

(1) 'On June 10, 1891, I saw a beautiful phenomenon—suddenly, at the zenith, E. of the Great Bear, shone forth a yellow globe like Venus at her brightest. Dropping somewhat slowly, it fell obliquely southwards. As it passed in its brilliant career it lighted up its dusky path with a glorious lustre. When it had descended about half-way down towards the horizon, it burst into a sparkling host of glorious fragments, each dazzlingly shot over with all the hues of the rainbow.'

(2) 'Date and time: 1892, December 12, 11h. 22m. G.M.T.

Object: Fine meteor, nearly $=2$.

Path: $55° + 41°$ to $45° + 20°$; length, $22°$.

Duration of flight: $1\cdot2$ second.

Colour: Bluish-white.

Appearance: Brightest in latter portion of its path, where it left a white streak for about one second.

Probable radiant: ϵ Ursæ Majoris.'

"If all such reports were modelled according to the latter plan the sufferings of the meteoric computer would vanish, for he would find himself in the presence of materials admitting of easy reduction and promising results of the highest value.

"Some years ago I remember receiving several accounts of a brilliant fireball, and, after comparing them, concluded that another description of the end point seemed necessary to clear up doubts. An early post fortunately brought me another letter referring to the same object, and as I eagerly read it, I thought, 'Ah, this is just what I wanted!' But the observer, in alluding to the end of the phenomenon, only said, 'at its disappearance the meteor got into a cloud, where it heaved one mighty flash, and all was over!' This was too cruel! Of course, it may be beneficial for amateur observers to relieve their pent-up feelings in this way, but they should have a thought for the agony occasioned to the poor would-be computer."

Steady persistent practice in noting these particulars will soon give the observer increased skill. One item requires especial attention—the duration of the meteor. Mr. Denning tells us that he has trained himself by observing the flight of arrows. He has employed a friend to shoot these to distances from fifty to two hundred yards at right angles to the line of sight, the elevation being varied as much as possible, and by repeating these experiments he has learned to judge intervals of from one to five seconds with an average error of less than one-fifth second.

All the above particulars, and not merely the direction of the paths alone, are of value in the determination of the radiant point. The meteors of one radiant have similar characters as to colour, streaks, etc., and also as to rate of motion. The apparent length of path is affected by the height of the radiant point, Mr. Denning noting of the Perseids of August 10 that, whilst between 9 and 10 o'clock in the evening the brighter meteors average a course of about 30°, in the morning hours when the radiant is near the meridian their paths are only one-third the length.

As in all good work, skill is not acquired at once, and the would-be meteor observer will find that he makes many failures to begin with. His first successes will probably be with some bright slow-moving meteor, and as these are relatively few, he will probably have to wait some considerable time before he can accomplish much. This need of patience and practice is one great reason no doubt why so few take up a pursuit which requires no equipment and which soon becomes full of fascination. Another is to be found in the unfortunate fact that from midnight to dawn is a much more fruitful time than from sunset to midnight, since the meteors which come to meet the earth are necessarily much more numerous than those that overtake it, and the earth has its sunrise point in front as it moves forward in its orbit, its sunset point behind.

Yet there are always prizes to be secured. There is a great pleasure when some brilliant wanderer flashes by in knowing that one has secured as full and accurate a record as possible of its appearance. It was seen but for a moment,

> " Like a snowflake on the river,
>   One moment white, then gone for ever."

Yet it has left something behind, something permanent,

something which years after may be eloquent of un-suspected truth.

The great Perseid shower, chief of all those which are of regular annual recurrence, has been rich in such indication. It has shown itself to be in intimate con-nection with the Third Comet of 1862 discovered by Swift. It has been traced night after night for a very considerable time before the date of its maximum, August 10, the radiant point travelling steadily back-ward in the sky from the borders of Cassiopeia and Andromeda in the middle of July to those of Camelo-pardus in the middle of August; the steady shift of the radiant, night after night, having been abundantly demonstrated by observations as well as being in strict accordance with theory.

The two most celebrated showers, after that of the Perseids, both fall due in the month of November; the Leonids, and the Andromedes. The history of the former shower goes back 1000 years to October 12, 902 A.D., a sufficient number of records being extant between this date and November 11, 1799, to show that the shower came in great force on an average three times in a century, and that the day of the shower was moving slowly onwards in the year. The astonishing display which took place on November 12, 1833, which from the accounts preserved would seem to have been the most impressive astronomical spectacle ever witnessed, enabled Prof. Newton and Prof. Adams to demonstrate that the shower was due to an immense swarm of meteors travelling in an elliptic orbit round the sun in a period of $33\frac{1}{4}$ years; while Prof. Schia-parelli showed that Tempel's Comet, 1866, I., moved in practically the same path.

The great shower of November 13, 1866, added much to our knowledge, and important but less abundant displays were seen in the two following years. After 1869 conspicuous showers from the radiant in Leo ceased, but trained meteor observers have hardly ever failed to notice a few characteristic meteors from this point of the heavens on November 14, or the nights immediately preceding and following.

As there appeared to be a slight increase in the number of meteors as early as 1896, public expectation of a repetition of the grand spectacles of 1833 and 1866 began to be excited in November, 1898, and the interest was increased the following year. It is matter of history that on neither occasion was there anything to answer expectation; a few Leonids indeed were seen, but nothing which by the utmost stretch of language could be described as a great shower. The reason of the failure is matter rather of conjecture than of knowledge, Dr. Johnstone Stoney and Dr. Downing considering that the orbit of the meteors has been so far perturbed that the main stream now passes clear or nearly clear of the earth's orbit.

The Andromedes, the second great shower of November, are in all respects a great contrast to the Leonids. The Leonid radiant does not rise on November 14 until 10.30 in the evening; the Andromede radiant is up the entire night, being nearly in the zenith when the Leonid radiant is rising. The Leonid meteors are extremely swift; the Andromedes are very slow. The Leonids are distinguished by their green colour, suggesting the presence of magnesium; the Andromedes are rather yellow, as if sodium were their chief constituent.

The history of the Andromedes is as well known as that of the Leonids. Whilst the latter approach the sun at their perihelion as nearly as the earth does, and recede somewhat beyond the orbit of Uranus at aphelion, the Andromedes only recede about half-way between the orbits of Jupiter and Saturn. Their period therefore is one of $6\frac{1}{2}$ years as compared with the $33\frac{1}{4}$ of the Leonids, and the greatest showers that we have had from them in recent years have been in 1872, 1885, and 1892. The shower of November 27, 1872, was one of peculiar interest, inasmuch as it was then clearly recognised that the swarm was moving along the same orbit which had been travelled by the lost comet of Biela, the comet which divided into two portions in December, 1846, and which has never been seen since 1852, when it returned, still in two portions.

Whilst the Leonid shower has been falling gradually later and later in the year, so that November 15 is now its date of maximum, the Andromedes, or Bielids as they are indifferently called, have moved from November 27 to November 23.

In sharp contrast with the apparent shifting of the position of the Perseid radiant during the months of July and August, referred to above, has been another fact which long years of patient work have enabled Mr. Denning to demonstrate—namely, the existence of radiants which do not shift, radiants which endure for many months together. Here was a circumstance which could not have been anticipated, which was indeed in flagrant contradiction to the theory of meteoric motion, and which even yet remains without any adequate explanation. Yet one single observer, by sheer patience and perseverance, has driven home

o

the unexpected, unexplained, seemingly impossible fact, and after having been long rejected even by experts, the fact of stationary radiants has at length received general recognition.

Such a fact, unexampled in the history of astronomy, ought to make many a meteor hunter. For six thousand years men stared at meteors and learnt nothing, for sixty years they have studied them and learnt much, and half of what we know has been taught us in half that time by the efforts of a single observer.

The following list of the principal meteor showers is reproduced from the *Knowledge Diary* for 1902, p. 72, and will prove fully sufficient for the beginner. An ampler list is supplied by Mr. Denning, year by year, in the *Companion to the Observatory.*

Name of Shower.	Radiant Point. R.A.	Dec.	Epoch.
Quadrantids	230	+53	Jan. 2—3.
κ Cygnids	295	+53	Jan. 14—20.
a Draconids ...	211	+69	Feb. 1—4.
a Serpentids ...	236	+11	Feb. 15—20.
β Leonids	175	+10	Mar. 4—15.
β Ursids	161	+58	Mar. 24.
Lyrids	270	+32	Apr. 19—21.
η Aquarids	338	− 2	Apr. 29—May 6.
η Pegasids	333	+27	May 29—June 4.
δ Cepheids	335	+57	June 10—28.
a-β Perseids ...	48	+43	July 23—Aug. 16.
δ Aquarids	339	−12	July 25—31.
Perseids	45	+57	Aug. 9—11.
γ Pegasids	5	+10	Aug. 25—Sept. 22.
β Piscids	346	+ 0	Sept. 3—8.
η Aurigids	73	+42	Sept. 12—Oct. 2.
Orionids	92	+15	Oct. 15—24.
δ Geminids... ...	106	+23	Oct. 14—29.
ζ Taurids	55	+ 9	Nov. 2—3.
Leonids	150	+23	Nov. 13—15.
ε Taurids	63	+22	Nov. 20—28.
Andromedes ...	25	+43	Nov. 23—24.
Geminids	108	+33	Dec. 1—14.

CHAPTER II.

THE ZODIACAL LIGHT.

THE earliest English description of the Zodiacal Light of which I know was given by Dr. Childrey at the end of his "Britannia Baconica,' published in 1660. It is as follows :—

> "There is a thing which I must needs recommend to the Observation of Mathematical Men, which is that in *February* and for a little before and a little after that Month (as I have observed several Years together), about 6 in the Evening, when the Twilight has almost deserted the Horizon, you shall see a plainly discernible way of the Twilight striking up towards the *Pleiades* or Seven Stars, and seeming almost to touch them. It is to be observed any clear Night. There is no such Way to be observed at any other time of the Year that I can perceive, nor any other Way at that time to be perceived darting up elsewhere. And I believe it hath . been and will be constantly visible at that time of the Year."

This description of the Zodiacal Light is quite sufficiently accurate for our ordinary English experience. In the tropics, however, it is seen far more constantly, and attains a brilliancy and distinctness of which we seldom have any example here. There, not only during spring, but more or less during the whole year, if the western sky be watched after sunset, as the twilight fades out, it will be seen that the twilight which at first seemed to be a pretty regular arch in the west, begins to show a tongue of somewhat greater brightness, which becomes clearer and clearer as the background of the sky around becomes darker, until at length it stands out defined as a great nebulous patch of light, broadest and brightest near the horizon and fading gradually away to the right and left and towards its apex. At its brightest part, and when best seen, it often much outshines the Milky Way; by as much perhaps as a couple of magnitudes—that is to say, about six times; *i.e.*, it is as much brighter than the Milky Way as the latter is in excess of the average brightness of the sky. But such a degree of brightness is confined quite to the centre of the light and to the portion nearest the sun; its borders melt indefinitely away until they are no brighter than the background of the sky.

The shape of the Zodiacal Light varies. It is broadest close to the horizon, where it may be as wide as 25° or even 30°, and tapers somewhat quickly at first. At 60° or 70° from the sun, it has become much narrower, and its edges, so far as they can be discerned, are nearly parallel.

It is easy to see why this beautiful and mysterious object is so much better seen in the tropics than in the temperate zone. The twilight is so much more prolonged in the latter; and the Light is of so elusive a character that a three days' old moon is sufficient to blot it out. It cannot, therefore, be seen here nearly so soon after sundown as in the tropics, partly because the ecliptic, with which its axis nearly coincides, is lower in our skies than in equatorial regions, and partly because our twilight is so much more prolonged.

If we take it that it is not until about an hour and a half to two hours after sunset that we can see the Light in this country, then at the end of February or the beginning of March we shall have the point of intersection of the ecliptic and equator upon the horizon just about the time when the Light is beginning to show itself. And, as the accompanying

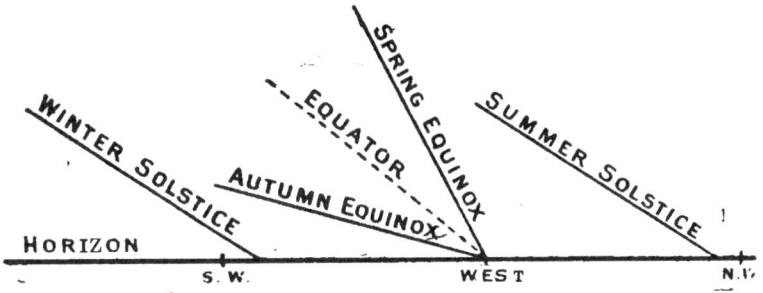

FIG. 34.—Inclination of the Ecliptic when the Equinoctial and Solstitial Points are on the West Horizon.

diagram will show, the angle which the ecliptic makes with the horizon is greatest at this time of the year; so that the Zodiacal Light rises up more abruptly into the sky than at any other time, and its brightness is therefore least affected by the absorption of the lowest strata of our atmosphere.

Although the Zodiacal Light has been more or less under observation for some three centuries—the great Kepler having carefully observed it, with the result of convincing himself that it was the atmosphere of the sun—the nature of the Light still remains more or less of a mystery. We do not know yet whether it lies in the plane of the ecliptic, or of the sun's equator, or between the two, or whether even its plane may not shift from time to time. It seems to vary in brightness, both according to the season of the year, and from one year to another, but the determinations of its brightness are usually far too vague and rough for any definite period to have been yet fixed for its changes.

So that we have in the Zodiacal Light the great anomaly of a vast astronomical object requiring no observatory and no telescope for its observation; and not only requiring none but permitting none; and yet to-day, when astronomy has lasted 5000 years, we are still in ignorance of many of the most fundamental facts respecting it.

This is due without doubt to the difficulties which attend its observation. Not that those difficulties are in the least insuperable, but they are very real. We will suppose that someone has noticed the Light for the first time and desires to make a record of what he sees. It at once strikes him that a mere eye-sketch of it is of very little good indeed; he must place it with respect to the stars. In all probability most of those which would be naturally used to define the outline of the Light are unfamiliar to him. He has therefore to have recourse to the star atlas. He painfully identifies the stars one by one, but each recourse to the atlas, which must necessarily be examined in the light, dazzles his eyes for his open-air work. He finds, therefore, that the process of recording what he has seen is a very slow and tedious one, and, dissatisfied with what he has done, speedily gives up the work. So that it happens that the names of the men who have done really useful work in this field may be counted almost on the fingers of one hand.

Yet this difficulty can be surmounted without much trouble. First of all, as I have already said, he who would become " an astronomer without a telescope " must learn his stars. They form the very alphabet of the language which he has to learn, and a little trouble spent here will soon repay itself. Next, the difficulty of recording his observations in the dark may be got over in several ways. It is possible to learn to write in the dark with sufficient clearness, and such little dodges as having sets of cards prepared, ruled with lines made by drawing a penknife across the back of the card, and cutting it partly but not entirely through, will be found helpful. Or the note book may be placed so that the rays from a ruby photographic lamp* may fall upon it. If the eyes are carefully screened from the direct light of the lamp, it will be found that the page may be lighted up quite sufficiently for the purpose of writing without the sensitiveness of the eye to the faint Zodiacal glow being much affected. If a chart is needed for comparison with the sky this might be made by tracing the map of the region required from some star atlas on a piece of thin cardboard and pricking little holes for the stars. A lamp can be used behind the card to show these, or a piece of card painted with luminous paint might be placed underneath. If a lamp is used, it will probably be found a convenience not to put the card in a vertical position before the lamp, but to construct such a box as is shown in the accompanying diagram. A glass plate forms the top of the box, and upon this the cardboard star-map lies. Within the box is a reflector of white cardboard, placed at an angle of 45° to the vertical, to reflect the light of the lamp up to the star-map. If a piece of tracing paper is now pinned down to the top of the box, the little holes

---

* A ruby lamp is mentioned as this kind is the easiest to procure, but a lamp of a deep green would probably be better.

representing the stars may still be seen, and the position of the Zodiacal Light, with respect to them, can easily be drawn. The places of three or four of the principal stars

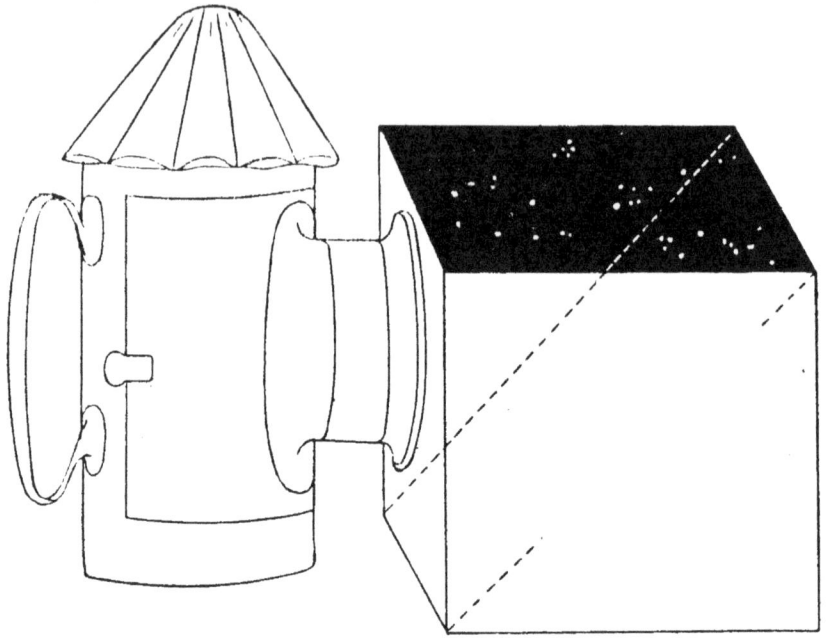

FIG. 35.—Illuminated Star Chart.

must be marked on the tracing paper to serve as reference points when it is taken off the box. Many similar dodges for getting over this initial difficulty will suggest themselves to those who seriously take up the work.

But, it will be objected, since the Zodiacal Light is seen so much better in the tropics than here, what is the use of trying to observe it in England? There is great use. Take for example one question ; the question of its variability in brightness from year to year. In a way this could be as definitely determined from observations made in England as from those made in any other single country. A careful record year by year for a term of years of the number of days when the atmospheric conditions were favourable, and when the Zodiacal Light was well seen, seen faintly, or not seen at all, would soon show as to whether there was any periodicity in its variation, and, if so, whether it varied with the sunspot cycle or not; just as Hofrath Schwabe's record of the days when the sun was seen to be free from spots in each year was quite as effective in determining the sunspot variation and the length of its period as exact measurements of the areas of all the spots would have been. In a certain sense our less favourable position would serve as a kind of photometer of the brightness of the Light, and our very hindrance might transform itself into a help.

Then, a more important point, observations in one latitude alone are not sufficient. We want to ascertain, either what is the amount of parallax which the Light shows or else that it has no perceptible parallax at all. Then, the degree to which its apparent outline is affected by atmospheric absorption is even more important, as otherwise we cannot tell whether an apparent shift in its plane is real or not. For both these enquiries it is necessary that observations should be made in several distinct latitudes.

The following are the principal points for observation in Zodiacal Light work. First of all, note the character of the evening. The magnitude of the faintest stars visible in the west should be recorded. The visibility of the Milky Way, and the distinctness with which its rifts and streamers can be made out, would be most useful for comparison. The evidence must be clear that there is no mist or dust veil to hinder observation, and here, it may be added, that the dwellers in towns are necessarily too severely handicapped to enter upon this

class of work. The smoky atmosphere and the glare of
street lights are fatal to so delicate a research.

The Light itself should then claim attention. It will
be perhaps easiest, first of all, to map out its extreme
border, and this will often be best detected by looking a
little way from the Light; "partially averted vision"
having a distinct advantage for very faint objects. The
position of the apex of the Light is very important,
and in the spring of the year it should be especially
noted whether the Light can be definitely traced beyond
the Pleiades. There can be no doubt that that group
does seem to exercise a strong attractive influence upon
the Zodiacal Light, probably apparent only, but on that
account the exact position of the apex relative to the
cluster is worthy of the very strictest attention.

The outline having been laid down, and the apex of the
Light having been fixed as carefully as possible, the axis
of the Light may next be located, and this should be done
independently of the charting of the border which has
just been finished. Several observers see a marked difference between the inner and outer portions of the Light.
If such a difference seems to be sufficiently clear, then the
outlines of both should be mapped down.

The Rev. George Jones, whose observations made in the
year 1853-5, when Chaplain of the United States steam-
frigate "Mississippi," are classic, not only always dis-
criminated between the inner brighter part of the Zodiacal
Cone, which he termed the "Stronger Light," and the
fainter outlying portion, which he called the "Diffuse
Light," but not infrequently noted close to the horizon
a region brighter than the Stronger Light, to which he
generally refers as the "Effulgence"; whilst he sometimes
saw the sky outside the Diffuse Light as "slightly paled,"
as if the Diffuse Light was continued further in a much

more attenuated degree. There were no sharp boundary lines between these different intensities of light; "the Stronger Light passed by degrees into the Diffuse, and the latter also gradually faded away. Yet there was in the former case a line of greater suddenness of transition," which could be made out without much difficulty; and so again in its degree with the outer boundary of the Diffuse Light.

This division of the Light into two, or even four regions, is an approximation to the method of observation recommended by Prof. Arthur Searle,. viz., the method of contour-lines. The latter the beginner will probably find difficult at first, though it is a more truly scientific mode of observation. In this method a portion of the sky or of the Milky Way is taken as a standard of brightness, and a line is drawn to represent as nearly as possible the positions where the Zodiacal Light is equal to this standard. This is work which needs much practice, for it is not possible to determine the position of a little piece of the line, then look away in order to mark it down on the chart, and then look back at the sky to get the next piece. The eye will need time for recovery after each turning to the chart. But when once some skill in the work has been acquired, the method of contour-lines will be found the most valuable. The extreme outline of the Light is in a sense a contour-line, as it marks the position of the faintest perceptible light; but the axis will not be a contour-line, for though it marks the line of greatest brightness, yet the axis is far brighter near the sun than at a great distance from it. Where the method of contour-lines is followed, it should not be confused by lines drawn on a different plan.

Beside the Zodiacal Light proper, or the " Zodiacal Cone," as it has been called from its shape, two other

related phenomena should be looked for in the neighbourhood of the ecliptic. One of these, a faint diffused light
which travels through the heavens in opposition to the
sun, has been called on that account the "Gegenschein"
or "Counterglow." It is somewhat elliptical in shape, and
about 7° in breadth. The centre and boundaries of the
Counterglow should be fixed with reference to neighbouring
stars as nearly as possible whenever the phenomenon is
seen, but its extreme faintness renders it a difficult object
to find at first; though once recognised, its position may
be followed with some degree of certainty. It attains its
greatest altitude at midnight at the winter solstice, but
cannot be observed then as it is interfered with by the
Galaxy. The end of January or beginning of February
is a more favourable season, as it is then passing through
the dull region of Cancer.

The "Zodiacal Band" is a fainter phenomenon still,
and much fainter than the Milky Way. It is a broad belt
lying along the ecliptic, and forming a prolongation of the
Zodiacal Cone. The Gegenschein is its brightest portion.

Besides these there appear to be very faint *permanent*
bands on or near the ecliptic, due to lines of very small
stars. One such band is noted by Prof. Arthur Searle as
having a breadth of about 2° or 3°, and lying south of
Beta and Eta Virginis, and extending from them nearly
to Spica.

Morning observations of the Zodiacal Light are even
more needed than the evening, for much fewer are made
of the western than of the eastern branch, and the one
series is requisite to complete the other. For reasons
analogous to those given above with respect to the eastern
or evening branch, the end of October is the best time for
looking for the western or morning branch. As a three-
day old moon is sufficient to kill the Light, from "the

dark of the moon" to near the full is the time for watching the morning branch; from a little after the full to two days after new moon for watching the evening branch. The question as to whether the moon causes any variation in the Zodiacal Light is almost an insuperable one, as the opportunities for observing it change with the progress of the lunation. The watch for the Light should therefore be always an item of the programme for a total eclipse of the moon.

The Rev. George Jones found that the moon itself was accompanied by a small glow analogous in appearance to the Zodiacal Light. This he found should be looked for a very few minutes before moonrise at the full of the moon, at times when the ecliptic makes a high angle with the horizon; *i.e.*, in February and March. Beside this "Lunar Zodiacal Light" he noticed on two occasions, when the moon was in her first quarter, and consequently still high in the sky at sundown, a "Luni-solar Zodiacal Light" due to the joint action of both sun and moon; a bright streak of light, that is to say, lying along the ecliptic, and visible even in the moonlight. The moon on one of these occasions was $4\frac{1}{2}°$ south of the ecliptic, but this streak of light lay along the ecliptic, pointing not to the moon, but to one side of it.

Such appearances need carefully looking for, and if seen, the most sedulous attention, to enable the observer to be sure that he is not being led away by imagination or by some purely local effect, quite disconnected from the actual Zodiacal Light. So, too, the pulsations noticed by Humboldt, Jones, and others should be watched for. The latter writes of them very confidently, and is convinced that they are not a mere physiological effect. He describes them thus:—"The changes were a swelling out laterally and upwards of the Zodiacal Light, with an increase of brightness in the Light itself; then, in a few minutes, a shrinking back of the boundaries, and a dimming of the Light; the latter to such a degree as to appear, at times, as if it was quite dying away; and so back and forth for about three-quarters of an hour; and then a change still higher upwards, to more permanent bounds."

The best atlas for the study of the Zodiacal Light is the "Atlas Cœlestis Eclipticus" of Heis, published at Cologne in 1878, and containing the zodiacal stars in eight charts, drawn to overlap each other widely. This atlas is not to be confounded with Heis's general star-atlas, the "Atlas Cœlestis Novus." The positions of the horizon-lines for different times and latitudes are marked on the borders of the charts, which will be found a great convenience.

Keen eyesight, patience, and a small star-atlas are, it will be seen, all the equipment that is required for Zodiacal Light work. The description of the work may not seem inviting, yet when once it is taken up, the looking for that strange, beautiful, yet faint and elusive glow will be found full of interest, and the more its peculiarities are followed up, the more will the sense of its mysteriousness be realized, and the greater will be the desire to contribute something which may explain its secret.

CHAPTER III.

Auroræ.

THE study of auroræ has made considerably more progress than that of the Zodiacal Light, and several striking facts relating to them have already received sufficient demonstration.

The points which have been established are of great importance. First of all, we know that though, strictly speaking, meteorological phenomena, auroræ have a close astronomical connection. They vary in number as observed in any given locality in accordance with the sunspot cycle. More than that, they are evidently in the closest sympathy with the disturbances which take place in terrestrial magnetism. They are practically non-occurrent in England at the sunspot minimum, but become more frequent as sunspots become more numerous, reaching a maximum about the time of the greatest solar activity.

Auroral observation demands, beside good eyesight, an observing station remote from the glare of towns and artificial lights. The stories are common enough of fire engines being turned out to quench an aurora, and, on the other hand, it has not seldom happened that a very mundane conflagration has passed muster for a "celestial display." "In the Memoirs of Baron Stock-mar an amusing anecdote is related of one Herr von Radowitz, who was given to making the most of easily picked up information. A friend of the Baron's went to an evening party near Frankfort, where he expected to meet Herr von Radowitz. On his way he saw a barn burning, stopped his carriage, assisted the people, and waited till the flames were nearly extinguished. When he arrived at his friend's house he found Herr von Radowitz, who had previously taken the party to the top of the building to see an aurora, dilating on terrestrial magnetism, electricity, and so forth. Radowitz asked Stockmar's friend, 'Have you seen the beautiful Aurora Borealis?' He replied, 'Certainly; I was there myself; it will soon be over.' An explanation followed as to the barn on fire. Radowitz was silent some ten minutes, then he took up his hat, and quietly disappeared."

Granted the suitable position the most important consideration for the student of auroræ to bear in mind is the absolute necessity for keeping as systematic a watch as possible. The general agreement between the cycles of sunspots, of magnetic variation and of auroræ is clearly established, but there are many questions arising as to the connection between their minor fluctuations. Now the observation of the magnetic elements is perfectly continuous. Self-recording magnets are set up at many observatories, and supply us year in and year out with an unbroken register. Our record of the state of the sun's surface is practically continuous also, but from the nature of the case auroræ cannot be presented in the same manner. The chronicle is broken by the intervention of cloudy nights. It is weighted by the difference in length of darkness between winter and summer. Further, it is difficult to express our auroral observations on a perfectly uniform numerical scale. One year may have a poor record, either because auroræ were actually rare, or because the observer was remiss or the weather unfortunate. Another year may present a fallacious appearance of abundance simply because the observer was more diligent or more lucky in the circumstances of his observations. In a word, the accidental errors of the work are large, and it therefore becomes the first duty of the student to keep his own personal part in the matter as systematic and as free from accident as he can.

This is the first essential, and the observer therefore should draw up a scheme for himself for the examination of the sky at certain definite hours, and for certain fixed intervals, to which he should adhere with the greatest possible regularity. There is no need for him to make any great inroad into the ordinary hours of rest, as the meteoric observer must do, or that his watches should be very prolonged. It will be sufficient if they are perfectly regular.

It is much to be desired that auroral observers should be scattered as widely as possible, that we may be able to present not merely the auroral conditions for a single place, but for the entire planet. It has already been discovered that auroræ are most frequent in two zones, one in the northern and one in the southern hemisphere, and that these zones shift their position with the progress of the cycle. In mid-latitudes, as in England, auroræ are most frequent at the time of the sunspot maximum. They retire polewards as the sunspot frequency declines, and are most frequent in high latitudes at the sunspot minimum. The place of the observer, therefore, is not a matter of indifference. A broken record in England cannot be pieced out by observations in the Shetlands or in Iceland.

But the value of a regular system of observations carried on at a single station for many successive years is very great, and we cannot have too many observers in the field.

Auroræ ·differ so widely in their character, that some sort of analysis of their leading forms is necessary in order to impart definiteness to records of them. The following scheme is due to the late Mr. J. Rand Capron, F.R.A.S., of Guildford* :—

1. The glow may be either entire or in patches, and yellow, red green or white, as the case may be, or with those colours mingled.

2. An arc or arcs, sometimes vertically, or multiple horizontally, forming a low bow or bows upon the horizon.

3. An evenly dark space between the arc and the horizon, considerably darker than the sky above the bow.

4. Bright rays, streamers, or coruscations, continuous or in groups, starting from above, and sometimes below, the arc.

5. Transverse, horizontal (or nearly so) streamers, or rays, crossing those thrown out from the arc.

6. Apart from the arc, streamers, or rays detached, or in groups horizontal or vertical, which latter, springing from E. and W., sometimes form arches uniting in a luminous point or ring, called the corona (effect of perspective).

After the mere fact of an auroral display has been recorded, the following details should be noted :—

1. The time and duration of the aurora.

2. The position and extent of the glow.

3. The position, height, and breadth of the arc or arcs.

4. The streamers or bright rays should be catalogued,

---

* *Astronomy for Amateurs*, p. 312.

as to their general appearance, positions, length, and as to whether they are curved or straight.

5. The colours throughout the aurora should be noted.

6. Any movements of the streamers and the direction of such movements.

7. The direction and speed of pulsations and their character; whether they are running sheets of continuous light, or finer patches, which become more luminous in rapid succession.

8. If the aurora is seen during a calm, sounds should be listened for, and if suspected, the observer should be most particular to ascertain if possible whether they can be due to any causes in his immediate neighbourhood.

An important detail in auroral work is the fixing of the position of some specially bright point from two or three fairly distant stations with a view to the determination of its height. This can obviously be best done by reference to the stars if many of these are visible at the time. It would, however, be well to have at hand some rough and ready means for obtaining the altitude and azimuth of any given point, and for this it would be easy to make a sort of rough wooden theodolite or altazimuth, with a bar carrying a big easily seen pair of sights upon it instead of the telescope. As the auroral flashes come and go so quickly the time of any such determination must be taken with jealous exactness.

The rough altazimuth described on pages 130 and 131, and shown in the diagram on page 133, would serve very well for this work, but the pin-hole tube should be replaced by a tube specially designed for auroral work. The eye-end of the tube should be closed with a cap, in the centre of which is a hole about half an inch in diameter; the object-end of the tube should be not less than one-

fourth the length of the tube in diameter, and it should
be crossed by two wires at right angles to each other.
These wires should be fixed in a collar so that they can

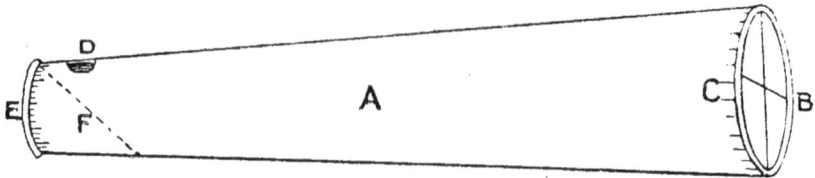

FIG. 36.—Tube for Observation of Auroræ.

A, tube; B, cross wires at object-end; C, graduated circle at object end; D,
eye-hole; E, graduated circle at eye-end; F, diagonal mirror in tube.

turn round in the tube, and the collar should be divided
into degrees so that the position angle of the wires may
be read off. As auroræ frequently attain great heights,
passing even through the zenith, it is a decided improve-
ment if, instead of having the eye-hole in the axis of the
tube, a small mirror is introduced and fixed at an angle of
45°, so as to reflect the light of the aurora up to a hole in
the side of the tube in the manner of a diagonal eye-piece.
The eye-hole and the mirror might also be made to
revolve, and in this case the eye-end also should be
marked with degrees, in the manner of a position circle.

The value of having some means always at hand, however rough, for determining the position of an auroral
beam, together with the need for exactness in giving
the time of the observation, was well illustrated by the
remarkable auroral beam of 1882, November 17. A
great sunspot, the largest visible for eleven years, was
nearing the central meridian of the solar disc. The
magnets, which had been uneasy from the time of the
first appearance of the spot at the east limb, began to
be seized with·the most violent convulsions about two
hours before noon on the 17th, the disturbance lasting

till 6 o'clock the following morning. "Strong earth currents were also observed at all the times of magnetic disturbance, varying in magnitude with the intensity of the magnetic changes, and the most violent electric storm recorded for more than thirty years swept over Europe and America." In sympathy with these manifestations a superb auroral display was witnessed on the evening of the 17th, but by far the most unique and striking phenomenon occurred "at about 6 p.m., when a bright beam of light rose from the eastern horizon and passed majestically across the sky in much the same manner as any ordinary celestial body might do, but with several hundred times their rapidity." Some twenty-six observations of the phenomenon were collected together by Mr. Rand Capron, but most of these were very incomplete, and their discussion was therefore attended with much difficulty; yet imperfect as the observations were they seemed to show with considerable probability that the height of the beam was 133 miles, and its speed about 10 miles per second. The direction of its flight was from east to west, magnetic not geographical. Had three or four of the observers but possessed some simple means for measuring the height of the beam at its culmination and the azimuths of its rising and setting, the precision of these conclusions would have been greatly increased.

The same charts that are useful for meteor observations may very conveniently be used for auroræ, the positions of the streamers or of the auroral crown being sketched in with reference to the stars. In all the work the first thing to be aimed at is to make the record as definite as possible. It is here that the difficulty of auroral observation is most felt. They are beautiful and impressive as spectacles, and the student will need no instruction in the preparation of his general descriptions. But to pick out the particular phenomena to which the desirable amount of definiteness can be ascribed will require practice.

From time to time curious beams of light are seen in the sky, the exact nature of which it is difficult to determine. Thus on March 4th, 1896, a curious light was seen stretching up from the horizon towards the Pleiades, which some observers were inclined to regard as auroral, some as the Zodiacal Light, and some actually regarded as being cometary. The fact that an unmistakable aurora was seen the same evening pointed strongly in favour of the auroral theory. On the other hand, as its direction coincided nearly if not precisely with that of the axis of the Zodiacal Light, and as similar beams have been seen in the same position on other occasions, the question cannot be regarded as absolutely decided. It would be a matter of the highest interest could it be shown that certain definite regions of the heavens were subject to recurrent flashes, and a careful collation of observations made at widely-separated stations would soon settle as to whether we should regard them as auroral or zodiacal, and could not fail to increase our comprehension of one or the other phenomenon.

CHAPTER IV.

The Milky Way.

THE short nights of midsummer do not in general give much opportunity to the Astronomer. But in summer and autumn the most wonderful of all celestial objects stretches itself across our English zenith and sweeps downwards to either horizon. This is that

> " Broad and ample road whose dust is gold
> And pavement stars, as stars to thee appear,
> Seen in the Galaxy, that Milky Way,
> Which nightly as a circling zone thou seest
> Powdered with stars."

Its sweep at midnight in mid-July is from the north-eastern horizon where the constellation Auriga is just rising, through Perseus and Cassiopeia on to Cygnus in the zenith; descending again on the other side through Aquila, Serpens, Sagittarius and Scorpio to the horizon in the south-west. It continues to cross the zenith at midnight until mid-December, when it sweeps upwards from the south-eastern horizon in Argo, between Orion and Gemini to the zenith now marked by the constellation Auriga; from whence it passes downwards through Perseus and Cassiopeia to the north-west horizon where the constellation Cygnus is setting.

The Galaxy is no modern discovery. Ptolemy of Alexandria has handed down to us a very full and precise description of it, and it has caught the attention and stirred the imagination of races even as savage as the Australian black fellows. It has been thought of as the roadway of the Gods by which they passed from their halls of eternal light, when they wished to visit this nether world of ours; or it is "Die Jakobsstrasse," the mystic ladder which the patriarch saw in his dream at Bethel, up and down which the angels moved.

Ptolemy and the Greek Astronomers had recognized two leading facts concerning it. One, that it marked out a zone in the sky, the centre of which was nearly a great circle; the other, that it was not equal and regular everywhere, but varied in different regions, in breadth, in brightness, in colour, in distinctness, and especially that in some places it broke up into two distinct streams. So much therefore was known about it long before the invention of the telescope, and though it gives to our greatest telescopes their most gorgeous starfields, though in some portions it still defies the efforts of our most powerful instruments fully to resolve it, though its characteristic formations are only brought out when we are dealing with stars far fainter than can be individually detected by the unaided sight, yet the Milky Way as a whole is essentially a naked-eye object.

The dwellers in cities and towns, smoke-veiled and flaring with arc lamps or incandescent lights, must abandon all hope of a really intimate knowledge with the delicate structure of the Milky Way. But there are many and many stations in this our island—lone country houses, little villages—upon which the stars of the short dark summer night will shine down like silver points set in ebony. The faint twilight, visible all night long above the northern horizon, will not interfere with the darkness of the zenith and the south. The evasive moon recognises that the season belongs of right to her more powerful brother, and either does not show herself at all, or timidly skirts the south as if anxious to escape notice. So though the summer hours of darkness are so few, sufficient of them may be utilized for so delicate a study as that of the Milky Way.

As the year advances the opportunities for the work increase. The nights lengthen and the Galaxy still crosses the zenith at midnight until December is far advanced. It is therefore very favourably placed for observation at almost any hour of the night during the months from August to November inclusive, and as we enter the new year the evening hours may still be devoted to it, though it passes into a less favourable position for the small hours of the morning. The worst time of the year for its observation is midnight at the end of March, and by consequence the late evening hours of April and May, and the early mornings of February and March.

The reason why it is so pre-eminently a naked-eye object is easily seen. The field even of a comet-seeker or any other telescope of wide field and low magnifying power deals with an inconsiderable fraction of the whole sky. It is impossible in the telescope to mark out the boundaries of the Way; to see where it radiates and divides; where it reunites and condenses again. It can only be examined piecemeal, a very small fraction

at a time—"the wood cannot be seen for the trees."
It is necessary, therefore, if we are to gain any
adequate knowledge of the structure of the Milky Way
as a whole, that we should supplement telescopic and
photographic examination by the most careful and
thorough scrutiny with the unassisted sight.

Fɪɢ. 37.—C. Easton.

This is astronomical work of a high order of importance which has been very seldom adequately attempted;
indeed, it is not too much to say that it was not until the
middle of the nineteenth century that it was first seriously
undertaken. A few distinguished names then occur as

having done good work in this field : Hermann Klein, Julius Schmidt, Heis, the astronomers of Cordoba under Gould, Trouvelot, Boeddicker and one or two others. More recently still Mr. C. Easton has brought out a beautiful and important monograph, illustrated by four large plates, entitled "La Voie Lactée dans l'Hemisphère Boreal," from which the accompanying map of the Milky Way in Cygnus is reproduced on a reduced scale. But valuable and important as are the charts which Boeddicker and Easton have given us, they cannot be considered as having attained finality.

It is not my intention in these papers, either to describe what other observers have seen or to give any regular history or summary of observations. That has been done most excellently before. Nor do I wish to describe what an observer might be expected to see for himself, since I fear in many cases the reader would content himself with the description. My intention is simply to give such merely general indications of the work which may be attempted and the manner in which it may be set about, that those who wish to do so may themselves undertake observations which, so far as they are concerned, may be original.

Given the absence of the moon, a suitable time of the year, and a thoroughly dark clear night, and even the most casual observer will at once perceive that the Milky Way is a most complex object. In one place we find it broad, and diffused ; in another it narrows almost to disappearance. Here the outline will be sharp ; there it is fringed out into faint filaments. In some places it coagulates into knots and streaks of light ; in others it is interrupted by channels of darkness. And amongst these I would specially invite attention to that region of which Gamma Cygni is the centre, and which extends from the borders of Cepheus to those of Aquila.   Here begins that great rift in the Galaxy the interpretation of which is so essential to

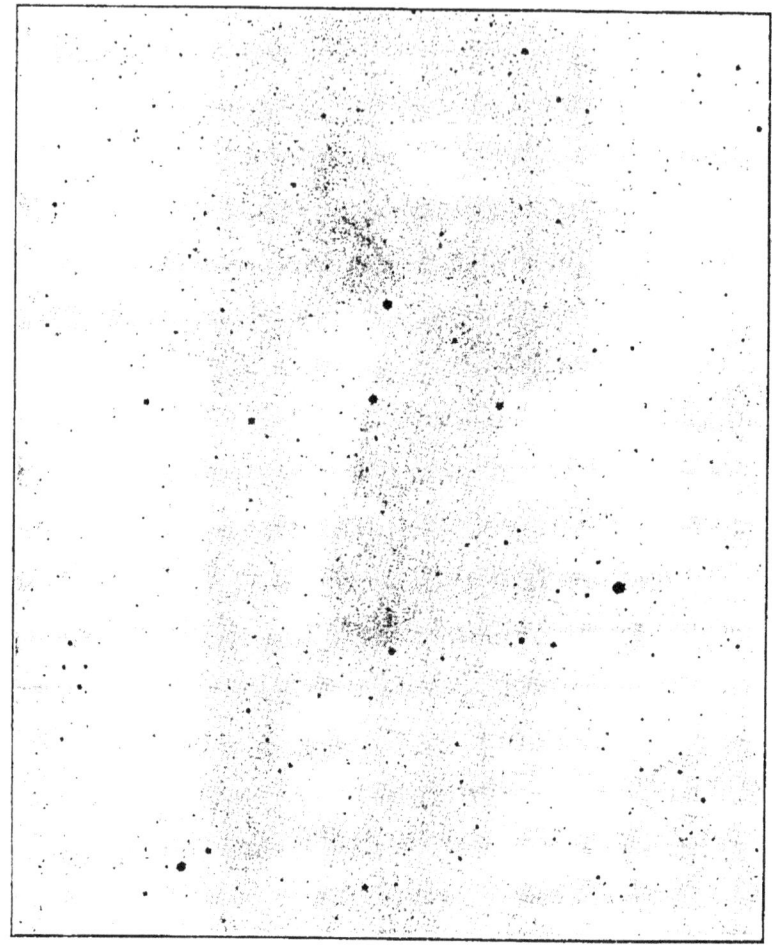

FIG. 38.—The Milky Way in Cygnus.
From Mr. C. Easton's " La Voie Lactée."

a true understanding of its meaning.   Here too are seen numerous crossways and side-rifts, not so easily caught as the main channel, but which will be detected as the observer gains experience and skill.

As to the actual method of observation, the first essential is that the observer should be screened from all interference by artificial light. Here comes in the same sort of difficulty that is experienced in drawing the Zodiacal Light, a difficulty to be overcome in much the same manner. First of all the observer must learn thoroughly the principal stars of the district which he is examining; then perhaps the easiest method is for him to dictate to an amanuensis, close at hand, but the light of whose lamp is perfectly shielded from the observer. The latter then might describe the course with respect to the leading stars, of the various rifts or rays, and at the same time should add estimations of the relative brightness or darkness of each respectively. Another method would be to carefully plot the stars down upon a sheet of paper beforehand, which paper might be illuminated by a very faint ruby light, like that used in a photographic dark-room; and the outlines might be drawn on the paper with reference to the stars by its means. The light itself must of course be arranged to shine only on the paper not on the observer's face. It is possible that a card covered by luminous paint might also be useful in this work, but it is not a device which I have myself employed, and I think it would probably dazzle much more in proportion to the amount of assistance it gave than would the faint ruby light. If the luminous paint is used, I should be inclined to recommend either that it be used under a card in which holes have been punched to represent the stars or under a sheet of ground glass or tracing paper or cloth on which the stars have been indicated by black dots.

It is the examination and representation of the details

of the Milky Way which form the work which requires to be done. Its general shape and direction are sufficiently well known. In particular the faint streamers and extensions which branch off from it, and some of which are very far reaching, need careful study.

Beside the detection of the shape of the various details of the Galaxy, their relative brightness needs to be recorded. This will be best done, no doubt, by a strictly comparative method. Let two areas be chosen, the positions of which have been first defined with reference to neighbouring stars, and record whether area a is brighter, equal to, or fainter than area b. Further, when experience has been gained in the work, it may be possible to make a numerical estimate of the difference in brightness between the two areas. Naturally the areas to be compared will be chosen of not widely differing brightness. In this case, if they are so near equality that one is only just perceptibly brighter than the other, or appears the brighter of the two a little more frequently than the other, as the observer turns his attention to and fro between them, the difference might be expressed as one step, and the observation might be written as follows:—

$$a > b, \; 1.$$

A slightly greater difference would be represented by 2, and so on; but it will be undesirable to make a comparison involving more than four or five steps. The comparison must not be made by attempting to bring the two areas into view at once, but by looking backward and forward several times from one to the other, that the same part of the retina may be employed for both. The time of the comparison must also be noted, so that the relative altitude of the two areas may be supplied later. So far as possible, of course, areas of about the same altitude will be compared together, as the effect of atmospheric absorption is exceedingly noticeable on faint objects.

As regards the kind of night upon which the Galaxy may be studied, Mr. Gemmill gives a hint well worth remembering: "The best views I have had of it have generally been in the breaks between showers, in the spaces separating heavy clouds drifting before a light wind." It is not at all necessary that the heavens should be clear from verge to verge. The work to be done is the examination of the details of the Galaxy, and portions of it are often displayed with unusual clearness under the above conditions. Mr. Gemmill adds: "A still dark night, following upon a windy and showery day, affords sometimes the best opportunities. But it is certain that the opportunities will be found—if watched for."

The beginner should bear in mind that though the Astronomer's rule is to note, that is to record, whatever shines *(quicquid nitet notandum),* nevertheless that he must learn to see before he can record. The careful study therefore of the chosen region of the Milky Way for two or three nights before any drawing is made will not be thrown away, and it should not be forgotten that faint lights are best seen, not from the centre of the eye but from the side, by "averted vision," that is to say. On the other hand, directly the observer feels that he is beginning to get some acquaintance with his subject he should begin to record. The first attempts will no doubt cost some effort, and may prove disappointing, but skill in delineation as well as in detection will come with practice.

CHAPTER V.

New Stars.

The appearance of a "new star" has, in all ages, been felt to be an impressive occurrence. The constellation groupings are so permanent in their character, that to see of a sudden some old familiar pattern amongst the stars changed in its features by the sudden appearance of a new member, a star like the other stars and not a planet, for its place undergoes no change—is so at variance with our ordinary experience, that it is no wonder that our forefathers regarded such an event as partaking of the supernatural. In the times before the telescope—indeed we might go further and say in the times before the spectroscope—such an event brought no information with it. It was impressive, it excited curiosity, but it conveyed scarcely any lesson. The spectroscopic examination of "new stars," on the other hand, has been extraordinarily fruitful, though we are very far as yet from being able to fathom the exact meaning of the facts which we have observed. One thing is clear, namely, that bodies appearing so suddenly as "new stars" have always done, and fading away again so quickly, must differ entirely from the great host of permanent stars. And yet we cannot but feel that the changes through which a "new star" may pass in a few weeks, and the order in which those changes succeed each other, may throw much light upon the changes which have marked in the past, or will mark in the future, the life-history of the more stable members of the heavenly host. It is this thought which makes the watch for " new stars" of such importance. They offer to us a key, which, however imperfect, is the only one which we can hope to find to unlock the secrets of stellar evolution. And that the "new star" may give us the fullest information within its power, it is essential that it be subjected to the scrutiny of the spectroscope whilst its light is still on the increase. The importance therefore of a stringent watch on the heavens does not lie at all in the *éclat* which will justly attach to the observer who is fortunate enough to be the first to detect a stellar outburst, but in the supreme importance that not one of the few short hours during which the star's light is on the upgrade may be unnecessarily wasted.

Such watching is not for the casual star-gazer, nor for the *dilettante* who has never taken the trouble to master the star-groupings and the coming and going of the planets. The planets especially are sad foes to such unqualified aspirants; and just as the "Crab" Nebula, Praesepe, and even the Pleiades, have sent many eager comet-hunters in hot haste to claim a comet medal, so Venus, Mars and Jupiter have inspired hundreds of letters to observatories or to newspapers to draw attention to "the wonderful new star." One of the most amusing instances of the kind was when the discoverer of the pseudo-planet Vulcan announced to the Paris Académie des Sciences his discovery of "a strange object in Leo," which proved to be no other than the planet Saturn.

The first duty, therefore, of the watcher for "new stars,"

is to work slowly, steadily, and systematically through the constellations till he has made himself thoroughly acquainted with the appearance, brightness, and position of their every member. When he has done this, then night after night it will be his task, whenever the skies are clear enough, to carefully scrutinise all the stars within his view. The labour will be great, but it must be borne in mind that such acquaintance with the heavens and such regular scrutiny of them are necessary for all the varied branches of " Astronomy without a Telescope," so that the watch for " new stars " may well be incidental only to other lines of work. To be the discoverer of a " new star " renders an observer rightly famous, but as the experience of the past fifty or sixty years shows that we cannot hope to record a discovery of the kind more frequently than once in nine years on the average, it is clear that it ought to be made subsidiary to some more fruitful line of research. The systematic eye-study of the Milky Way is particularly a form of astronomy which might be combined with it; since it is on the Milky Way, or on its branches, that nearly all the *Novæ* have been found, as the accompanying diagram (Fig. 39) will show.

The most famous of all " new stars " is, of course, the one which appeared in the constellation Cassiopeia in November, 1572, and which is always associated with the name of Tycho Brahe, since, though he was not actually the first to discover it, he has left us the fullest and most systematic observations of it. It was lost to sight in March, 1574, after having been visible for seventeen months. Thirty years later another "new star" appeared, only less famous than the *Nova* of Cassiopeia. This one was also observed for seventeen months, and is always associated with the name of Kepler, though its actual discoverer was not Kepler himself, but one of his pupils,

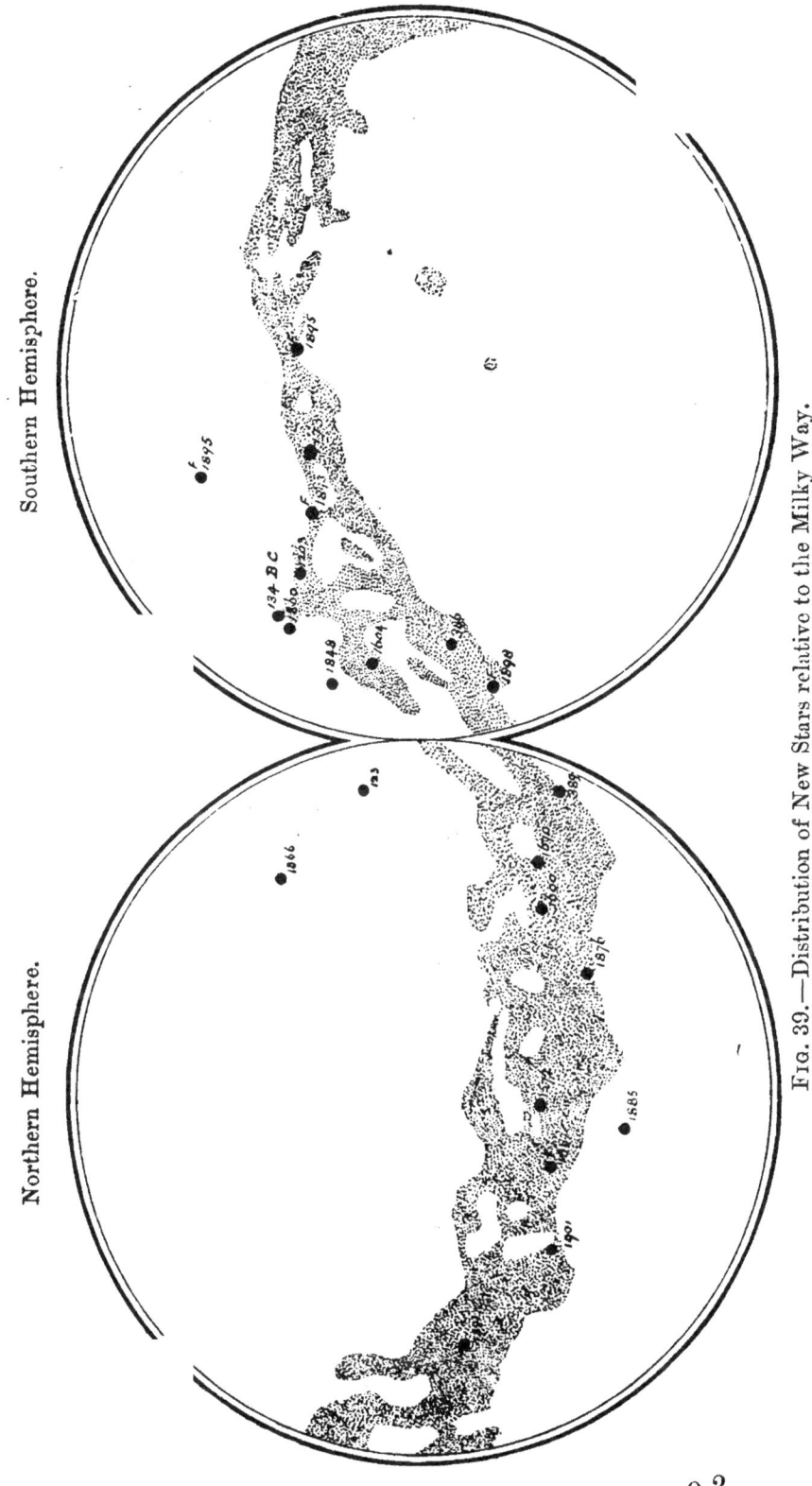

Southern Hemisphere.

Northern Hemisphere.

Fig. 39.—Distribution of New Stars relative to the Milky Way.

(The Novæ discovered by Mrs. Fleming on the Harvard Photographic Plates are distinguished by the letter F above the date.)

John Bronowski. Its position was in the right foot of Ophiuchus.

No such brilliant *Novæ* as these have been seen in more recent times, since both stars were reckoned, when first seen, to be brighter than Jupiter; indeed the star of 1572 ranked as equal to Venus when at her greatest brilliancy. The first of those noted in modern times was, like that of 1604, discovered in the constellation Ophiuchus. It was detected by Mr. Hind on April 28th, 1848, and was then increasing in brightness. Four days later it attained its maximum, which only ranked it of the fifth magnitude. The second star of the list was discovered in the globular cluster 80 Messier, which lies between Alpha and Beta Scorpii. The third was detected by Mr. Birmingham at Tuam on May 12th, 1866, when it was recorded as of the second magnitude. It may be doubted, however, whether this star is quite of the same order as the others. It had been observed several years previously as of magnitude $9\frac{1}{2}$, and it still remains visible, being now classed as a variable under the name T Coronæ. Still its outburst must have been very sudden, for Dr. Julius Schmidt, whose acquaintance with the heavens was of the very closest, declared that about two and a half hours previous to Mr. Birmingham's discovery, he had the constellation of the Northern Crown under his observation, but had noted nothing unusual. The next discovery was to fall to his own lot. On November 24th, 1876, he found a third magnitude star had appeared in the constellation Cygnus. This star rapidly faded away, and was only of the fifth magnitude on November 30th. Its spectroscopic history was of intense interest, for, by September, 1877, the light coming from the star was almost entirely monochromatic, and corresponded with that which would be given by a planetary nebula.

The next "new star," discovered independently by a considerable number of observers in August, 1885, was actually involved in a nebula, the great nebula of Andromeda, and was close to the nucleus. This was wholly a telescopic *Nova,* and since that time several other telescopic *Novæ* have been discovered by means of the photographic charts which have been made at the Harvard College Observatory, or at its southern annexe at Arequipa, Peru. But two *Novæ* have been detected besides these, both of which were naked-eye objects, and the history of their discovery is in the highest degree encouraging and instructive for the "astronomer without a telescope." On February 1st, 1892, an anonymous post-card was received by Dr. Copeland at the Royal Observatory, Edinburgh, with the following announcement:—

"Nova in Auriga. In Milky Way, about two degrees south of χ Aurigæ, preceding 26 Aurigæ. Fifth magnitude, slightly brighter than χ ."

That night the star was examined by means of an eye-piece prism at the Edinburgh Observatory, and its spectrum was seen to contain many vivid bright lines, which at once marked it as a "new star" of spectroscopic interest, not inferior to the one which had made so great a sensation in 1866. And the anticipations which that first glance gave were much more than fulfilled. Indeed, Nova Aurigæ opened an entirely new chapter in the spectroscopic study of stars of its class ; but such study is apart from our present purpose, which cannot be better fulfilled than by quoting the account given by the discoverer, the Rev. Thomas D. Anderson, in a letter which appeared in *Nature,* February 18th, 1892.

"Prof. Copeland has suggested to me that as I am the writer of the anonymous post-card mentioned by you a fortnight ago (p. 325), I should tell your readers what I know about the Nova.

" It was visible as a star of the fifth magnitude certainly for two or three days, very probably even for a week, before Prof. Copeland received my post-card. I am almost certain that at 2 o'clock on the morning of Sunday, the 24th ult , I saw a fifth magnitude star making a very large obtuse angle with β Tauri and χ Aurigæ, and I am positive that I saw it at least twice subsequently during that night. Unfortunately, on each occasion I mistook it for 26 Aurigæ, merely remarking to myself that 26 was a much brighter star than I used to think it. It was only on the morning of Sunday, the 31st ult., that I satisfied myself that it was a strange body. On each occasion of my seeing it, it was brighter than χ. How long before the 24th ult. it was visible to the naked eye I cannot tell, as it was many months since I had looked minutely at that region of the heavens.

" You might also allow me to state, for the benefit of your readers, that my case is one that can afford encouragement to even the humblest of amateurs. My knowledge of the technicalities of astronomy, unfortunately, is of the most meagre description ; all the means at my disposal on the morning of the 31st ult., when I made sure that a strange body was present in the sky, were Klein's ' Star Atlas ' and a small pocket telescope which magnifies ten times."

An examination which Prof. Pickering had made of photographs which had been taken at the Harvard College Observatory of the region of the Nova, showed " that the star was fainter than the 11th magnitude on November 2nd, 1891, than the 6th magnitude on December 1st, and that it was increasing rapidly on December 10th." It would seem to have attained a maximum about December 20th, when its magnitude was 4·4. It then decreased slightly for about a month, fading to somewhat below the 5th magnitude. When Dr. Anderson detected it, it was again on the upgrade, and it seems to have reached its maximum about February 3rd, after which it declined. The star therefore had been visible to the naked eye for fully six weeks before Dr. Anderson's discovery, a time when it was evidently passing through the most interesting and important changes. But for his zeal in studying the heavens, it would without doubt have escaped notice altogether, and the spectroscopic revelations which it yielded would have been wholly lost.

This success naturally stirred Dr. Anderson up to making the search for " new stars" his serious business, as he narrates in a most interesting letter addressed to the *Observatory* for March, 1902. He writes:—

"I need hardly say that before the advent of Nova Aurigæ my astronomisings were fruitless—fruitless, that is to say, so far as the rest of humanity was concerned—but far from being fruitless as regarded myself, for there was for me, at least, a certain joyful calm when, after a long evening spent in writing sermons or in other work, I threw up the window, and taking out my little pocket telescope, surveyed the never palling glory of the midnight sky. But after the appearance of Nova Aurigæ the thought occurred to me that perhaps after all 'new stars' might not be such rare phenomena as had up to that time been supposed. The correctness of that surmise has been proved by Mrs. Fleming's discovery since then of no fewer than five of these objects on the Harvard College photographs, although it is certainly strange that all these *Novæ* should have appeared in the southern hemisphere. I therefore resolved to commence a search for new stars."

Dr. Anderson's purpose was not confined to stars visible to the naked eye, but extended to all stars included in the great work of Argelander and Schönfeld. This necessitated his making charts for himself for the regions south of $+40°$, a work which meant the plotting down from their catalogue places of more than 70,000 stars. The instruments with which he worked were a large binocular and two refractors, one of $2\frac{1}{4}$ inches, the other of 3 inches aperture, and which respectively enabled him to see stars of the 10th and 11th magnitude. To continue his own account :—

"Thus armed I began to hunt for 'new stars.' I worked with

might and main, never going to rest as long as the sky remained clear, and often rising in the night to see if the clouds had passed away, and if they had, hurrying downstairs to begin work either with the binocular or with the telescope. The chief obstacle that I have to contend with in such work is that the only windows in this house from which I can thoroughly examine the heavens, face the north-west. Not only is my field of labour thereby very greatly circumscribed, my telescope being only able to command that part of the heavens which extends from the Equator to $+70°$, but the discomfort is frequently not inconsiderable, as the northerly and north-westerly winds, which so often bring with them transparent and unclouded skies, are in winter and early spring far from being balmy, and can make themselves felt even when the window shutters are partially closed.

"At first my search was mainly for *Novæ*, and was prosecuted by means of my binocular. . . . When I came to see that hunting for *Novæ* was not attended by the success which I had anticipated, I began, without entirely abandoning such work, to make a systematic search for variable stars. For this I used my $2\frac{1}{4}$-inch telescope, comparing what it showed me with the representation of the heavens contained in the B.D. charts. . . . . I was always glad if after three or four months of searching, during which I might have examined perhaps 20,000 stars and suspected fifty or sixty of variability, I was able at last to come across one whose brightness changed.

"I found Nova Persei, I need hardly say, without either binocular or telescope when I was casting a casual glance round the heavens."

It was then no mere happy chance that the Council of the Royal Astronomical Society were honouring, but the most persistent and strenuous work, when at the annual meeting of February, 1902, they conferred upon Dr. Anderson the Jackson-Gwilt Medal. The words of the President to him, when presenting the medal, put the case briefly and clearly.

"Nova Aurigæ was discovered by you on February 1st, 1892, when of the 4th magnitude, and but for your discovery it might have escaped observation. Nova Persei was discovered on February

22nd of last year at 2.40 a.m., when of 2·7 magnitude and low down in the sky. This early discovery of yours made it possible for Pickering to obtain its spectrum before its maximum was reached. It is no small matter to have discovered one of these *Novæ*, but it is a veritable *tour de force*, such as *à priori* would have seemed impossible to have discovered both, and I am delighted that we have the opportunity to congratulate you on your success and to do honour to your astronomical zeal and intimate knowledge of the sky."

One point with regard to the discovery of Nova Persei deserves further mention, namely that Dr. Anderson's discovery was made almost simultaneously with the outburst, for photographic records .show that so late as February 19th the star must have been fainter than the 11th magnitude. It had probably only entered the ranks of stars visible to the naked eye a very few hours when Dr. Anderson remarked it. But it is gratifying to remark that whilst, but for Dr. Anderson, Nova Aurigæ would have passed entirely without detection, the closer watch which is now kept upon the sky resulted in several entirely independent discoveries of Nova Persei. Herr F. Grimmler discovered it the same morning at Erlangen; Captain P. B. Molesworth at Trincomali was not quite twelve hours after Dr. Anderson, and four hours later still Mr. Ivo Carr Gregg, at St. Leonards, had also detected it, and communicating his observation to Col. E. E. Markwick, the Director of the Variable Star Section of the British Astronomical Association, the other members of that Section were made aware of the event before the news of Dr. Anderson's discovery had reached them through the ordinary channels of information.

CHAPTER VI.

The Structure of Comets.

In the chapter on "Morning and Evening Stars," I pointed out that the systematic observation of heliacal risings and settings offered a chance—a rare one, it is true, but still one not to be despised—of making the first discovery of a comet. Unfortunately comets, bright enough to be visible to the naked eye, have been but very scarce visitants, nor can we reasonably expect that they will be more numerous in the future. Still, when one does come, it justly attracts universal attention; and the "astronomer without a telescope" will naturally be anxious to know if there is any work within his power to effect upon it.

There is. He cannot of course expect to make useful determinations of the comet's place; nor can he scrutinise the changes which take place in the minute details of its head. But the shape, extent, and precise form and position of the comet's tail are better observed by the naked eye than with the telescope; since the eye can embrace a far wider field, and is the fitter instrument for dealing with great extensions of faint light. To map out, night by night, the precise position of the tail or tails with reference to the neighbouring stars, to trace its limit and to determine its exact form, are by no means unimportant tasks.

And for this reason. The last thirty years have seen the development and gradual acceptance of a theory which explains the origin and structure of those far-stretching wisps of light which our forefathers found so mysterious and awe-inspiring. The first step towards the elaboration of this theory was made by Olbers nearly a century ago in

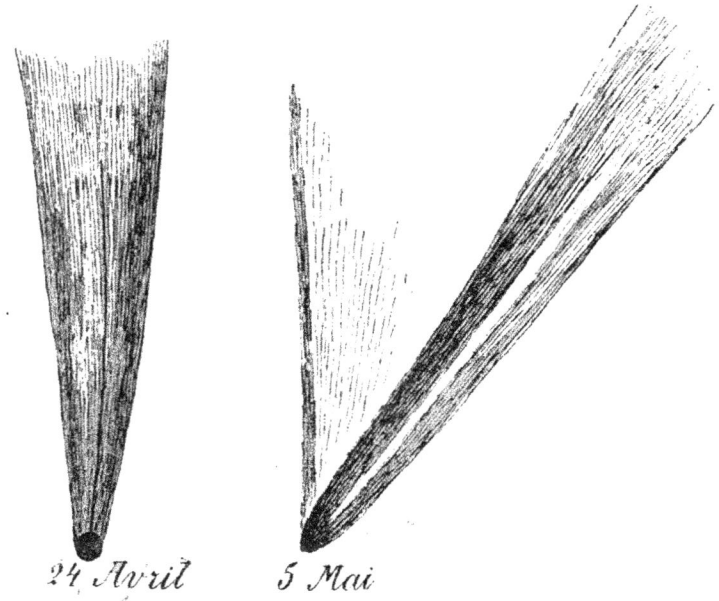

FIG. 40.—The Comet of 1901 on April 24 and May 5.

a memoir on the great comet of 1811; but in its present shape we owe it to Prof. Brédikhine, lately the Director of the Poulkova Observatory.

It was very early noticed that the tails of comets are in general directed away from the sun, and the instance of certain comets, which passed at perihelion very close to the solar surface, was sufficient to prove that we must not regard a comet's tail as forming a body coherent with the head. Thus the great comet of 1843 swept round some

180° of longitude at perihelion in something like eighteen
hours of time. The tail which had been seen before
perihelion, pointing away from the sun in one direction,
could not possibly have been composed of the same
material as made up the tail lying in the opposite direction
after perihelion. But if it were supposed that the sun

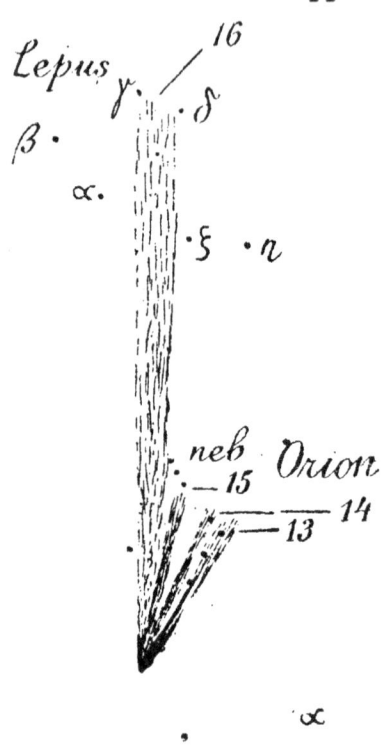

12 *Mai*

Fig. 41.—The Comet of 1901 on May 12.

were capable of exercising a repulsive force upon some
portion of the substance of the comet, driving it off in a
continuous stream, then the general behaviour of cometary
tails would be accounted for. The tail, seen at any
particular time, would be the summation of particles which
had left the comet at different successive instants, just as
the trail of smoke from the funnel of a locomotive, as seen

at any particular moment, is composed of particles that came off from it at successive instants, and is not a body coherent with the engine.

Such a repulsive force we may find in an electrical action of the sun; an action the efficiency of which would depend upon the surface of the body acted upon, as contrasted with that of gravitation, which depends upon its mass. Thus whilst the nucleus is moving in its orbit round the sun under the influence of gravitation, very minute particles in the envelope will find themselves practically less strongly attracted towards the sun, or it may actually be repelled from it; in any case the effect is to separate them more and more from the main body of the comet.

Of the particles thus driven away from the head, the lightest would be those most strongly repelled; and Prof. Brédikhine found that several of the great comets of the past century were distinguished by the possession of long straight tails which must have been composed of particles moving under an influence some twelve or fourteen times that of gravity. These long straight tails form an exceedingly characteristic class, and have accordingly been ranked by Prof. Brédikhine as composing his first type. Such tails are obviously composed of the lightest material entering into the composition of the comet, and assuming this to be hydrogen, then the average tail of the second type might well be made up of hydrocarbons; whilst iron and the heavier metals would, from their molecular weight, be suitable elements to form the third type. In the tails of the second type the effective repulsive force does not differ greatly from gravity, and it is this type of tail, slightly, but not extravagantly curved, which is most commonly observed in comets visible to the naked eye. The third type is usually sharply curved, short and brush-like, and the repulsive force is considerably less than that of gravity.

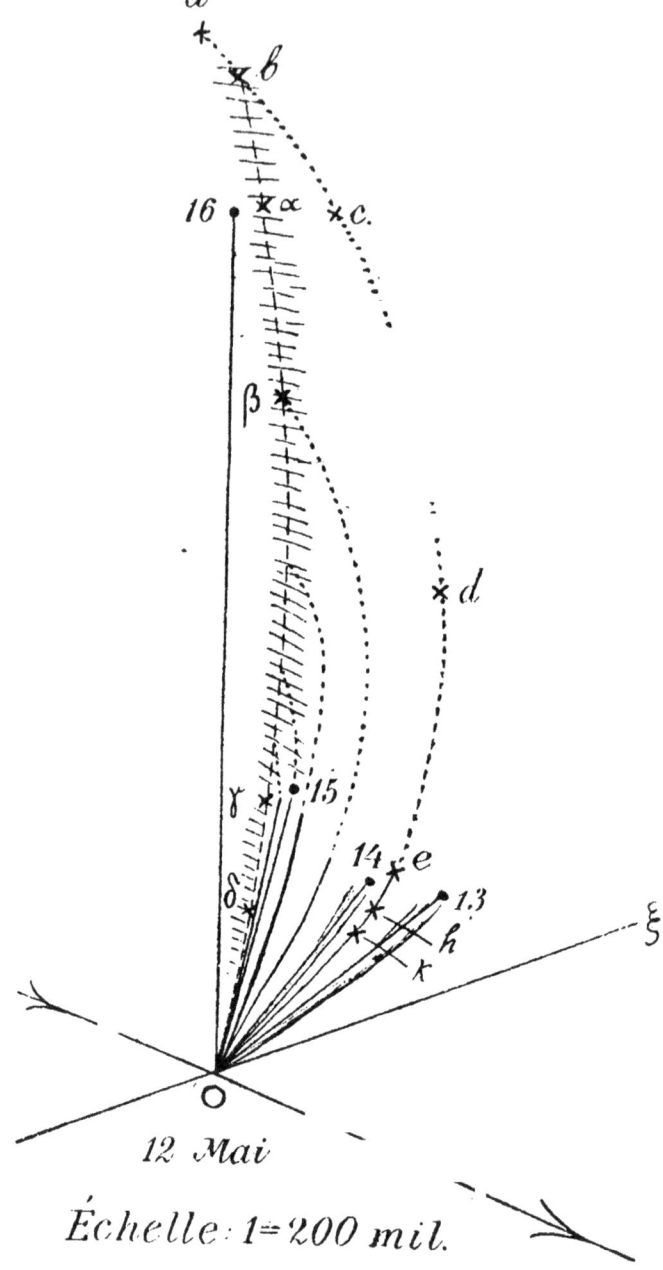

Fig. 42.—Prof. Brédikhine's analysis of the tail of the Comet on 1901, May 12.

The great comet seen in the southern hemisphere in April and May, 1901, has recently been analysed by Prof. Brédikhine, in a paper from which the three accompanying diagrams have been reproduced, and his results afford an interesting example of his method of treatment, and of the use which can be made of careful naked-eye observations of the positions of a comet's tail.

On April 24th, before the perihelion passage, the comet showed practically only a single tail, and that was of the first or hydrogen type. After perihelion, the tails were only of the second and third types, the matter composing the first type tail having been apparently completely driven away. On May 5th, the chief tail, which was distinguished by a very well marked rift, showed this rift as of a conical shape, the apex of which was occupied by the nucleus, and not as usual of a conoidal form. A drawing made by Mr. J. Lunt at the Royal Observatory, Cape of Good Hope, on May 12th, was of especial interest. It not only showed the principal tail of May 5th with its dark rift, but a long broad faint tail some 25° in length, and a short tail between the two. Prof. Brédikhine's analysis of this drawing is given in the accompanying diagram. The lines 13 and 14, the two branches of the principal tail, were due to a repulsive force a little greater than unity; the line 15 is due to a force of about 0·65; the broad faint tail, Prof. Brédikhine ascribes to substances of the third type, and finds that when the particles which make it up are traced back to the nucleus they indicate that a great explosion took place on April 22nd; a vast quantity of matter of a wide range of density being driven off in a single short-lived convulsion. The points in the diagram, a, β, γ and δ, correspond respectively to values of the repulsive force of 0·85, 0·65, 0·25 and 0·15.

CHAPTER VII.

A TOTAL SOLAR ECLIPSE.

IT is the misfortune of the British Isles to be so completely shunned by total eclipses of the sun that the last visible in England happened as long ago as 1724, and this country will not be favoured with the next until 1927. Yet the ease of modern travel brings the phenomenon within the reach of so many that it may be well worth while to glance at the various kinds of work which can be undertaken by "astronomers without telescopes," especially as in 1905 a total solar eclipse of considerable duration will be visible as near the British Isles as the northern provinces of Spain.

It should be borne in mind by all who are favoured with a good view of so rare a phenomenon as that of a total eclipse of the sun, that there is a kind of moral obligation upon them not to let the opportunity pass entirely without profit. "The giddy pleasure of the eye" is no excuse for selfishness. Each one should do something, make some record, which may hereafter be of service to others in the solution of some of the problems which an eclipse presents. We owe an inestimable debt to those who preceded us who did leave

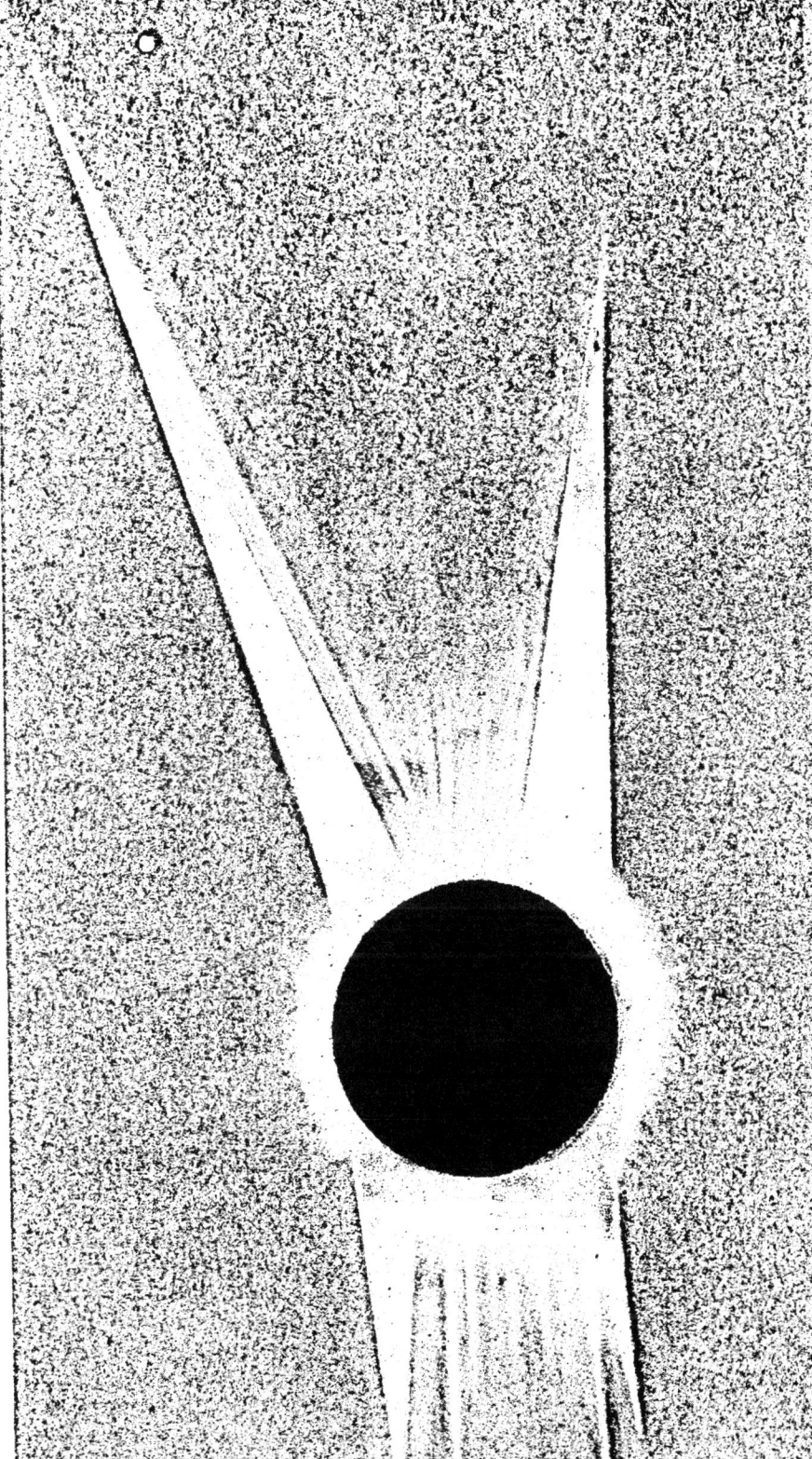

such records, and we can only repay that debt, by in like manner doing our best to leave material as useful for the benefit of those who shall in their turn succeed to us.

First of all, the most obvious work for anyone to undertake, who watches a total eclipse without a telescope, is to draw the corona. This may seem a very trivial matter, and when the strange discrepancies between different sketches are noted, a very useless thing to attempt, especially in view of the entrance of photography into the field. But it is not so. The chief fact that we have as yet established with regard to the corona is that it varies in form and character with the sun-spot cycle, and this fact, though supported by the photographs, was demonstrated by the comparison of drawings. Then again the careful examination of drawings has shown them to be far more trustworthy than a cursory look would suggest. The wide differences between different sketches have often been due to the sketchers choosing different sections of the corona; one choosing the brightest inner corona, another the fainter and more irregular contour, a third the faintest extensions. The results have really not been contradictory but rather supplemental of each other. Nor has photography entirely superseded the work of the sketcher even yet. The coronal streamers, often shown in drawings, were photographed in the eclipse of 1898 for the first time. The previous failure of photographs to record them had occasioned their very existence to be denied in some quarters, and had cast unmerited suspicion upon the drawings which delineated them.

The work of drawing the corona is, however, not one to be done off-hand. The intending artist should be one who has already acquired skill and quickness in draughtmanship. The time of an eclipse is terribly short, and the object to be sketched is bizarre and unfamiliar. There should be frequent practices before-hand, either upon drawings of the corona, held at a distance of 107 times the diameter of the eclipsing moon, or, perhaps better, upon little wisps of cirrus cloud. But in any case the time from the first sight of the object to the completion of the sketch must be rigidly confined to the time of the expected duration of totality. Quickness to see and record is the first essential for coronal sketching.

The next point to be noted is the need for fiducial lines by which to orientate the drawing. This may be done by providing a plumb-line right across the line of sight. If the weight at the end of the plumb-line dips into water, it will serve to steady it against vibration with the wind.

If several sketchers can combine they should portion out the corona between them before the eclipse begins, the vertical line being adopted as one of the dividing lines, and if four workers are present a line parallel to the horizon might be another, thus giving each observer a quadrant. A fifth observer might make a rapid out-line of the entire corona as a basis for combining the four quadrantal sketches.

The sketchers should be careful to indicate as precisely as possible the positions of any red prominences, as these can be verified either from photographs or from observations with the spectroscope. Distances from the limb of the dark moon should be carefully estimated in terms of its diameter.

In some former eclipses, notably in 1878, the brightest inner corona has been screened off by means of a black disc, so as to leave the eye more sensitive for the detection of faint coronal streamers. This is not recommended as it is a troublesome and very doubtfully useful device. But all intending sketchers should be most careful to avoid dazzling · their eyes during the coming on of the partial phase, and should rest them as much as possible shortly before totality.

Fig. 44.—Drawing the Corona; Buxar, India, 1898, January 22, showing the use of the plumb-line.*

(From a Photograph by Miss Gertrude Bacon.)

White chalk on purplish blue paper is an admirable material for representing the corona. Notes as to any colour or colours perceived in the corona should be made.

Quite another class of work may be taken up by those who have keen eyesight, in the search for stars. To note

* From "The Indian Eclipse, 1898"; report of the Expeditions organised by the British Astronomical Association.

which stars are seen, when they are first glimpsed, and when lost, would be of some value as a register of the clearness of the sky, and of the brightness of the eclipse, as well as for comparison with the records of old eclipses wherein the appearance of stars was observed.

The Zodiacal Light should be looked for, for though the chances against seeing it are very great, a single clear record of its appearance during an eclipse would be of the utmost value, and might decide at once whether its axis coincided with the ecliptic or with the solar equator.

The observations of the "shadow-bands" is one of some interest, and as it only requires a white surface and a few light rods, there should never be any difficulty in enlisting observers.

It must be remembered that the bands are usually very faint, and have to be definitely looked for. A white surface must be prepared to receive them; either a white sheet, which may be fastened down to the ground, or set up vertically on an upright frame, or a whitened wall. The surface should be marked with two black concentric circles of known diameters, that the intervals apart of the bands may be correctly judged. A rod should be placed to mark the direction of the bands themselves, as seen at the beginning of totality; and another to mark their direction of motion; another pair being used for a similar purpose for the bands seen at the close of totality; and after totality is over the most careful determinations must be made of the directions of the rods, and of the position of the sheet or wall. The following questions drawn up by Mr. E. W. Johnson for the assistance of the Members of the British Astronomical Association should be answered.

QUESTIONS.

1. How long before totality did the bands appear?

2. What number of bands were visible, say, in 10 seconds?

3. What was the direction of motion?

4. Were they inclined to the direction of motion?

5. What was the direction and force of the wind?

6. Did they come uniformly, or in batches?

7. What was their speed?

8. What was the width of the bands?

9. What was the distance apart of the bands?

10. Were they very faint, or clearly defined?

11. Was their direction after totality the same as before?

12. How long after were they visible?

13. Did you see any bands during totality?

The subject of shadow-bands leads naturally to meteorological work, for there is no doubt that the direction of the wind affects the direction of the motion of the bands. The meteorological observer should therefore provide himself with some form of vane and some means of ascertaining the force and speed of the wind. The wet and dry bulb thermometers would seem to be the next most important instruments to take, that the change in temperature and in saturation of the air might be marked. The barometer would come in the third place. It is of course desirable that observations should be made at regular intervals for some days both preceding and following the eclipse, especially at the same hour of the day as that when the eclipse takes place.

Those who find themselves about to witness an eclipse,

yet without any instruments or any preparations for observing, should not let it pass without some record. They should note the appearance of the sweep of the shadow over the country as it comes and as it goes; the colours of sky, land and sea should also be noted, the sky being divided into three regions—namely, overhead, at sun-height, and near horizon.

Photographic cameras come very close to a definition of a telescope, and hence should be excluded from the scope of the present paper. Yet as in all probability there are some hundreds of possessors of cameras for every one who possesses an astronomical telescope, it is perhaps not superfluous to remind photographers that a very large field is open to them. Cameras with a focal length of two feet and upwards may be profitably used upon the corona itself. In this case the camera should be firmly fixed and exposures not exceeding half a second should be given. If the focal length be not more than 15 times the aperture, this with an "ordinary" plate will probably be found quite sufficient. For shorter focal lengths shorter exposures should be used.

Hand cameras may be profitably employed for photographing the landscape during the approach and recession of the shadow. A series of photographs taken at five minutes intervals with a uniform speed of shutter, such as Miss Bacon took at Buxar in India, would give a very interesting and certainly very pretty record of the increase in the darkness as the eclipse comes on.

Finally, a valuable record of the total light of the eclipse can be obtained by exposing a plate in a printing frame to the light of the corona during totality. Portions of the plate can be exposed for different lengths of time or the plate itself may be placed under some form of sensitometer. Further information also would be obtained by using different coloured screens in connection with plates of varied colour sensitiveness.

The most successful attempt to sketch the corona without optical assistance, and the one organized on the largest scale, was that arranged by Members of the British Astronomical Association in the eclipse of 1900. By the permission of the Council of the Association, of Mr. H. Keatley Moore, and of Miss C. O. Stevens, I am enabled to reproduce here Mr. Keatley Moore's integrated drawing from the whole of the series of sketches made by Members of the Association on that occasion (*see* Frontispiece), and Miss Stevens' separate sketch made in the city of Algiers. For a detailed description of these drawings I would refer the reader to the account of the eclipse published by the British Astronomical Association, "The Total Solar Eclipse of 1900."

CHAPTER VIII.

Stars by Daylight; and the Sum of Starlight.

Are the stars visible to ordinary sight in the daytime? There is a widespread tradition that they are; that if an observer places himself at the bottom of any deep shaft—as of a mine, a well, or a factory chimney—which may shut off scattered light and reduce the area of sky illumination acting on the retina, he will be able to discern the brighter stars without difficulty. The tradition is one of a respectable antiquity, for Aristotle refers to persons seeing stars in daylight when looking out from caverns or subterranean reservoirs, and Pliny ascribes to deep wells a similar power of rendering visible the stars the light of which would otherwise be lost in the overpowering splendour of the solar rays.

The tradition, well founded or not, has often been adopted for literary effect. It seems almost sacrilegious to hint that no star known to astronomers could have shone down unceasingly upon poor Stephen Blackpool during his seven days and nights of agony at the bottom of the Old Hell Shaft; that at best he could only have caught a glimpse of it for a few minutes in each twenty-four hours as it passed across the zenith. Dickens indeed does not absolutely say that Stephen watched the star by daylight. It is only a natural inference from his description; but Kipling adopts the tradition in its extremest form when he writes of:—

"The gorge that shows the stars at noonday clear."

But is the tradition true? Of course everyone knows that Venus from time to time may be seen even at high noon; but then Venus at her brightest is many times over brighter than Sirius. Then, again, the assistance of a telescope enables the brighter stars to be discerned at midday; but the telescope not only directs the eye and greatly limits the area from which the sky light reaches the observer, but it enormously increases the brightness of the star relative to that sky illumination. The naked-eye observation of true stars in full sunlight stands in quite a different category.

Humboldt, who was much interested in the question, repeatedly tried the experiment in mines, both in Siberia and in America, and not only failed himself ever to detect a star, but never came across anyone who had succeeded. Much more recently an American astronomer set up a tube for the express purpose of seeing the Pleiades by daylight, also with no effect. It has been supposed that Flamsteed, the first Astronomer Royal, sank a well at Greenwich Observatory for the purpose of observing Gamma Draconis, the zenith star of Greenwich, in this manner. The existence of the well is undoubted, though Sir George Airy, the late Astronomer Royal, was unable to find it, but Flamsteed marks it on more than one of his plans of the Observatory, and there is a drawing extant of the well itself, showing the spiral staircase that ran down it. But its purpose seems to have been, not to have furnished the means of observing the star with the naked eye, but to enable the observer to measure telescopically as accurately as possible the distance of the star from the true zenith at the moment of transit.

Sir John Herschel mentions a case, which he considers as satisfactory evidence, of an optician who stated that the earliest circumstance that drew his attention to astronomy "was the regular appearance at a certain hour for several successive days, of a considerable star through the shaft of a chimney." This, it will be noticed, is second-hand evidence. I have never been able to obtain evidence even so direct as this myself, though I have met several persons who felt quite confident that they had seen stars by daylight on looking up the shaft of a mine, or that " some one had told them he had done so."

But the value of such indefinite statements is *nil*. I have met evidence more direct and explicit in support of that favourite legend due to the fertile imagination of the Emperor Jahangir, of the Indian juggler who threw a rope into the air and climbed up it, to disappear at its top into space—a legend which still makes periodic reappearances, and finds not a few devout believers. But direct, first-hand, scientific testimony of an observer who has been enabled by the use of a shaft to detect stars with the naked eye at midday is still to seek. By scientific testimony, I mean the record of the day, hour and minute when the star was seen, the latitude of the place, the depth of the shaft, the breadth of its mouth—the numerical elements, in a word. which are necessary to give value to the observations. There must be not a few of the many who take an interest in astronomy to whom the means for making such an observation are available, and who, if they would take the trouble, could report, "I have seen such a star at such a time," or, "I have watched for such a star at the time of its transit across the zenith on so many occasions, when the sky was clear, and could see nothing." Such observations would set the question at rest whichever way they tended; but what are wanted, here as elsewhere, are definite observations, carefully made, fully and systematically recorded; not vague, second-hand impressions which are perfectly valueless as evidence.

Whether or no the use of a shaft to diminish the effect of sky illumination, and so to render the stars visible by daylight is practicable, it suggests a method for dealing with what Prof. Newcomb in a recent paper has justly described "as among the most important fundamental constants of astrophysics," namely, the value of the total light of all the stars. In the paper* alluded to, Prof. Newcomb points out that the "total amount of light received from all the stars may serve as a control on theories on the structure of the universe, because the amount of light resulting from any theory should agree with the observed amount. It is also a quantity which we must regard as remaining constant from age to age." Yet, strangely enough, very few attempts have been made to determine it. One of these was made by Mr. Gavin J. Burns,† his method being to compare the brightness of the spurious disc of a star seen out of focus in a telescope with the light of the sky. The eyepiece of the telescope was pushed in and out until the brightness of the spurious disc seemed to correspond with that of the sky. Prof. Newcomb, two years later, adopted several plans, his purpose being a twofold one—first to determine the relative brightness of different portions of the sky, and next to express the brightness of given units of surfaces in terms

* *Astrophysical Journal*, December, 1901.

† *Journal Brit. Astr. Assoc.*, Vol. XII., p. 212.

of star magnitude, from whence in turn the brightness of the whole heavens in terms of starlight may be inferred.

Prof. Newcomb's first plan was to use a small tube, the length of which he could easily vary, the ends of the tube being covered with caps having apertures of varying diameters, and to measure therewith the smallest area of sky which was certainly visible to him. A second method was by means of small mirrors arranged so as to enable different regions of the sky to be compared directly. Roughly speaking, the Galaxy appeared, surface for surface, about twice as bright as the sky outside it.

For the determination of the brightness of different areas in terms of starlight a concave lens was used, so as to spread out the image of the star into a disc, and the brightness of the expanded image was cut down by means of an absorbing glass to that of the sky. The results appeared to point to a value for the total starlight of from 600 to 800 stars of magnitude 0, whilst Mr. Gavin Burns, by different methods, fixed the value at about 400 stars for one hemisphere, or 800 for the entire heavens. Both results, though more accordant than might have been expected, can only be regarded as first approximations, and there is abundant room for many other observers to follow these pioneers, and supplement their work.

One method for comparing the light of the sky in two different regions would be by means of some such simple apparatus as the following :—A tube, bent at right angles, should be fitted at the angle with a piece of card placed at 45° to either arm. The card should be painted a dead black all except a white cross in the centre. The observer should look down one arm, through a diaphragm about $\frac{1}{5}$ inch in diameter, and view the card, which would be illuminated by the light coming down the other arm, and the opening of which would be directed to some known region in the sky. This opening should be provided with a series of caps having apertures of different diameter, and the arm itself should be fitted with a draw tube, so that both the size of the opening and its distance from the card might be varied at will. The observer, having carefully set the tube in some given direction, would move the draw tube in or out, or vary the caps over the aperture, until the white cross on the cardboard in the angle could just be certainly discerned. The aperture, the length of the draw tube, and the part of the sky to which it is directed, must then be carefully recorded.

The rough altazimuth, described in the chapter on "The Sun and the Seasons,"* would prove a suitable mount for such an instrument. If used for this purpose its circles must be read, whilst, of course, the time of the observation should be taken, and the state of the sky noted. Necessarily, observations of this kind are only possible at stations far from the glare of towns, and on moonless nights of special clearness.

The observer might well begin his work with some such device as this, but in a field so nearly new there would be full scope for his best ingenuity and contrivance in improving on this beginning, and in arranging for better and exacter methods for dealing with the noble problem he had undertaken.

It must be noted that the result of these observations will give the sum of starlight + any other general source of illumination which may be present. It must be assumed that the observer is working far from the influence of any artificial lights, and that so far as he can ascertain there is

---

* *See* p. 133.

absolutely no cloud or mist in the sky. But there still remains the question whether the general illumination of the sky does not vary from time to time. Thus, two observers of the very first rank, Mr. Denning and Mr. Backhouse, have recorded that in August, 1880, the sky was unusually light. It is clear that the sum total of starlight cannot vary from time to time, and if the light of the sky is different on one occasion from that which it is on another, allowance, of course, being made for any annual variation, due to the Milky Way or some of the brighter constellations being in an especially favourable position, then this variation in luminosity must be due to some cause other than starlight. Over and above, therefore, the two very important researches,—(*a*) of the relative brightness of different portions of the heavens, and (*b*) of the total sum of starlight,—there will come the question as to whether there is in addition any variable source of luminosity, and, if so, what are its nature and origin, and the laws and causes of its changes.

CHAPTER IX.

Various Sky Effects.

THERE are certain phenomena which lie very near the
border line of astronomy and meteorology; so near that it
is difficult to say which science has the stronger claim to
take note of them. Amongst these, perhaps those which
have the best claim to be included in the department
of astronomy are the strange bright clouds which were
discovered by Ceraski, and which were afterwards made
the object of careful study by O. Jesse. These Luminous
Night Clouds were utterly unlike any phenomenon which
had been previously recorded, and their discovery, like
that of the Gegenschein, was a striking evidence that not
even yet have the fields of work which lie close at hand
been all explored. There is still an ample harvest to be
reaped in more than one direction by the man who can
reinforce an observant eye by thoughtful patience.

These luminous clouds were not visible at every time
of the year, but only during the nights of summer; their

period of visibility for Berlin, where Herr Jesse observed them, being from May 23rd to August 11th; a period corresponding nearly to the season of continual twilight for that latitude. Their light was derived from the sun, which during that season is never more than 18° below the northern horizon, their great height above the surface of the earth enabling them to catch his rays; for the comparison of photographs taken of them from different stations showed that they ranged from fifty to fifty-four miles in elevation, or ten times the height attained by light cirrus clouds.

In appearance these night-clouds were of a brilliant silvery-whiteness, slightly tinted at times with blue if near the zenith, or with a reddish-yellow tinge if near the horizon. They were woolly and striated in character, and repaid examination with a field-glass of large aperture, by means of which they might be traced considerably further than the naked eye could follow them.

The discovery of an order of clouds at a height above the earth so greatly exceeding anything which had ever been observed, even of the lightest cirrus, was remarkable enough. More remarkable still were their variations. For they were not by any means a permanent phenomenon, and diminished in frequency of appearance from the time of their first discovery. From 1885 to 1889, they were seen before midnight; later they could only be detected in the morning hours. Their movements were more interesting still, and were such as might be caused—so it has been suggested—if, though travelling with the earth, they were but lightly subject to its attraction, and experienced some retardation as they travelled with it.

From any point of view the existence of these clouds must be regarded as most remarkable. That clouds should exist at all at a height greater than the highest stratum to which we owe twilight, and that so existing they should be an occasional and variable phenomenon are entirely unexpected discoveries, and still remain unexplained. Can it be that they are one, of the by-products of the great volcanic eruption of Krakatoa in 1883 ? If so, they may be looked for after any great series of volcanic outbursts, such as that which commenced in May, 1902, with the destruction of St. Pierre in Martinique, even though these eruptions cannot compare in violence with that of Krakatoa.

Three striking sky effects followed that great eruption of 1883. The first was comparatively restricted both as to area and time, and took the form of a remarkable coloration of both sun and moon. At Batavia, in Ceylon, at various places in India, the sun was seen to be blue or green ; blue when at the zenith, changing through green and yellow to total obscuration near the horizon. A much more lasting effect was that which received the name of " Bishop's Ring," having been first reported from Honolulu by the Rev. S. E. Bishop on September 5th, 1883. This ring was a remarkable species of halo to be seen on every fine day surrounding the sun from its rise to its setting, and even occasionally round the moon. Thus in the Stonyhurst Observatory report for 1883 it is stated : " During the day the sun is invariably surrounded by an intense silvery brightness slightly tinged with green, and at a distance of about 20° from the sun this tint sometimes changes gradually into a pink or pale violet, and fades away at about 45° . . . . an orange tinted haze extending about 45° from the moon was also seen on several nights towards the middle of December." There can be no doubt that " Bishop's Ring " was a diffraction effect due to an immense quantity of dust particles of an extreme minute-

s 2

ness driven up by the great explosion to a great height in our atmosphere and slowly subsiding.   For as late as 1887 the ring still remained, though it could then be only traced as a peculiar white haze to a distance of about 10°.

The third result of the eruption of Krakatoa was the occurrence of "Afterglows" at sunset and "Foreglows" at sunrise.   These were distinguished from normal sunsets and sunrises in that they differed from them in the time of their appearance and the place or quarter in which they were formed; in their periodic action or behaviour; in the nature of the glow, which was both intense and yet lustreless; in the regularity of their colouring; in the colours themselves, which were impure and not of the spectrum; and, lastly, in the texture of the coloured surfaces, which were neither distinct cloud of recognised make nor yet translucent mediums.   The regularity of their colouring was particularly striking.

> " Four colours in particular have been noticeable in these after-glows, and in a fixed order of time and place—orange, lowest and nearest the sundown; above this, and broader, green; above this, broader still, a variable red, ending in being crimson; above this, a faint lilac.   The lilac disappears; the green deepens, spreads, and encroaches on the orange; and the red deepens, spreads, and encroaches on the green, till at last one red, varying downwards from crimson to scarlet, or orange, fills the west and south."*

These magnificent afterglows reappeared, but on a diminished scale, after the Martinique eruption of May, 1902.   A valuable letter by Prof. A. S. Herschel appeared in *Nature* for July 24th, 1902, describing the afterglow of June 26th.   The sun set about 8h. 25m., and a quarter of

---

* Gerard Hopkins in *Nature*, January 3rd, 1884, p. 223.

an hour later a long low belt of sky in the N.W. had grown orange-yellow, whilst the ruddiness of the sky in the east had by the same time risen nearly to the zenith. Between the two there lay a white tract about 30° in width, which was gradually invaded and at last quite occupied by the advancing ruddy colour from the east. About 8h. 55m. from the zenith down to 30° above the place of sunset, and for 40° or 50° on either side of the vertical line through it, was a broad expanse of rich rose-coloured, lake-red light. This red glow sank rather rapidly in height, and by 9 .p.m. it had ceased to be independently visible.

The colour effects of an ordinary sunset are due to the depth of atmosphere through which the sunlight reflected to us from the clouds has passed before illuminating them. In their passage the rays of short wave-length suffer a greater scattering, and are therefore lost to us in a greater proportion than those of greater wave-length, with the effect of producing a golden or ruddy glow. The afterglow would appear to be due to the intervention of a reflective stratum at a greater height than that of ordinary clouds In 1883 this was no doubt composed of the finely-divided volcanic dust to which " Bishop's Ring" had been due, suspended high in our atmosphere, and no doubt the afterglows of 1902 owed their origin to a similar cause. The height of that stratum can be inferred from the depression of the sun below the horizon at the time when the further boundary of the glow is setting, or is at some definite altitude, or from the elevation of the zones of its greatest brightness. In this way Prof. Herschel found the elevation of the reflective stratum on June 26th as about five miles ; whilst on June 28th, he estimated "the sun's parting illumination of the sky to rosy colour" at not much more than 30 minutes after sunset, and the height of the dust stratum in consequence as about $7\frac{1}{4}$ miles.

These intervals were very distinctly shorter than those remarked in 1883 and 1884. The primary glow after Krakatoa averaged 54 minutes after sunset until its disappearance on the horizon; the secondary averaged 96 minutes. If the secondary glow be due to direct sunlight, the average height of the stratum causing it must have been nearly 40 miles. But if, as appears more probable, the secondary glow was due to the same stratum as that which produced the primary, being in fact a second reflection from it, then the mean value of its height would be about 11 miles.

On a bright clear evening, as the sun goes down, an interesting phenomenon may often be watched. If the observer turns his back on the sun, he will see in the east, immediately after sunset, a long dark line spread along the horizon. This darkness, which is indeed "the Shadow of the Earth," thrown upwards as the sun goes down, rises somewhat rapidly, its upper edge, under favourable circumstances, being quite sharply defined. This "shadow of the earth," which may be made out more or less distinctly on any clear evening in a suitable locality, and is especially easy to watch at sea, revealed itself in a very interesting manner in the case of the luminous night clouds described in the beginning of this chapter. If the clouds were seen in the earlier portion of the night, that is to say before midnight, then the shadow of the earth covering them little by little would darken them from the top downwards. On the other hand, if they were first seen after midnight, then it was the upper edge that lighted up first, the cloud rising out of the shadow.

The "Earth's Shadow" came strongly into evidence in

the case of the Krakatoa "afterglows." The Rev. S. E. Bishop, writing to *Nature*, Vol. XXIX., p. 549, says:—

"I beg special attention to my former remark of the 'earth-shadow sharply cutting off' the upper rim of the first glow. This was very manifest in the strong heavy glows of September, showing clearly that the first glow directly reflected the sun's rays, while in the afterglow, which had no defined upper rim but continued much longer, the haze reflects only the light of the first glow."

And, again, in *Nature*, Vol. XXX., p. 194, he writes:—

"In your issue of April 10th (p. 549), is the statement by an observer in Australia that the 'red glow was margined by an immense black bow stretching across from the north-west to south-east.'

"I wish to say that the above language almost exactly describes the appearance to which I alluded on the same page as 'the earth-shadow cutting off the upper rim of the glow.' The 'black bow' of the Australian was evidently the shadow of the horizon projected on the haze stratum. In both the above cases the lower surface of the haze was evidently well defined, so that, as the horizon inter-cepted the direct rays of the sun, a well-marked shadow moved westward and downward. Above this black rim or bow appeared the secondary glow, produced by the reflection of the sun's rays from that portion of the haze surface which was directly illuminated. Very often the second glow was more conspicuous and impressive than the first, because it shone against the dark sky of night."

If the secondary glow were due to the reflection of direct sunlight, then no doubt its upper edge would have been sharply marked off by the earth's shadow, just as was the case with the primary glow. But its diffusion as compared with the definiteness of the earlier glow points to its being a reflection of the latter, a view strengthened by the fact that the depression of the sun below the horizon at the sinking of the second glow was as nearly as could be determined double what it was at the sinking of the first.

Prof. A. Riccò, in addition to these two points, mentions a third. From time to time in fine weather the phenomenon is presented at sunset of "Crepuscular Rays." These are broad pink ribands of light diverging like the sticks of a fan from the point where the sun has just gone down. They are formed in a very similar manner to the rays of light which we sometimes see in moisture-laden air, when the sun itself is hidden behind a dense cloud, and which children are accustomed to speak of as "the sun drinking." In the latter case, these bands of light are due to the sun shining out from between irregularities of the clouds and lighting up the laden air, which shines in the path of the rays just as the particles of dust do—"the motes in a sunbeam"—when light is admitted through a small aperture into a darkened room. In the case of the crepuscular rays, mountain ridges or banks of cloud may serve to partially intercept from the upper atmosphere the light of the sun which has set, throwing their shadows in long lines upwards, whilst the sunset glow shines out in the intervals between the shadows, and is seen by us as a radiating system of broad pink streamers. Here in England it is rare that these streamers can be traced very far towards the zenith; they are usually lost as they pass overhead, though sometimes their termination may be seen as counter rays in the east. Col. E. E. Markwick,[*] however, records that he has seen the rays in South Africa night after night extending from horizon to horizon; the radiation from the west and convergence again towards the east being, of course, an effect of perspective. In the case of the Krakatoa sunsets, Prof. Riccò remarked on several occasions these crepuscular rays as being conspicuous in the primary glow, but they were not seen in the secondary, a further proof that the former was due to direct, and the latter to reflected sunlight.

---

[*] KNOWLEDGE, 1901, April, p. 88.

If we look across water towards the setting sun, we see a broad track of light extending from below the sun's place towards us, due to the reflection of the sun from the surface of the waves. It occasionally happens that an analogous reflection is produced in the air from the under surface of ice-films floating horizontally. In this case a vertical shaft of light is produced, rising from the sun as a base, which is known as a "Sun-Pillar," and of which a very fine example was seen on March 6th, 1902, over the greater part of southern England. The difference between the formation of such a sun-pillar and the Krakatoa afterglows depends partly upon the greater elevation of the stratum giving rise to the glows, and partly because of the smaller size of the particles forming it. The ordinary atmospheric particles are very small compared with the wave-length of light, and hence scatter especially the shorter rays, those producing the sensation of blue. The particles of the Krakatoa dust were large in comparison with these, and hence scattered rather the rays of long wave-length—the red rays. The composite effect of the glows therefore resulted from the interaction of these two different orders of particles upon the rays passing through them. And over and above the richer and more complicated sunset colours that were thus produced, there was the reflection of the sunset glow itself by a stratum of highly reflective dust particles at a great elevation.

" Thus we may probably conclude that the haze which followed the eruption of Krakatoa, and produced the twilight glows, was composed mainly of very fine dust, and that this dust at a great altitude reflected the light of the setting or rising sun after diffraction through the stratum and diffraction and absorption by the lower atmosphere, and secondarily again reflected this reflected light."*

The heavens are the province of the astronomer, the atmosphere of the meteorologist, and all the various phenomena that have been referred to above belong to the atmosphere, and hence meteorology may be said to have a claim upon them. But they are distinguished from purely meteorological objects—such as the various orders of clouds, rainbows, haloes, parhelia and the like—in that they are connected directly with the earth's rotation, and with its position relatively to the sun. They are phenomena of the earth as a planet rather than of the earth considered as a world complete in itself, and from this point of view may be considered as belonging to astronomy. But they are referred to here, not so much on this account as from the illustration they afford of the value of the habit of exact observation. "Science," it has been often said, "is measurement"; it certainly depends upon the record of phenomena in numerical form. The difference in value is immense between the most vivid and picturesque description of a flight of a meteor and the half-dozen numbers which give the time of its appearance, its brightness, and duration, and the precise position of its path. It is only the latter which are of permanent value, and it is only from the habit of registering the obvious facts connected with a given phenomenon that the faculty is developed of recognizing other points needing numerical expression and record. Thus in the case of the beautiful sun-pillar of March 6th, 1902, though many vivid descriptions were written of it, so far as I know only one observer noted what should have been apparent to all, namely that the pillar moved an

---

* Hon. Rollo Russell: "The Eruption of Krakatoa," p. 195.

appreciable distance in azimuth, following the unseen sun in its northward movement below the horizon. So in the case of the afterglows it was only those observers who had made specific observations of ordinary sunsets, and who knew accurately their characteristic features, who could state definitely and with precision that the sunsets after Krakatoa were abnormal, and define wherein their peculiarities consisted. Striking as they were, it was by no means everyone who saw anything in them out of the common, and even so experienced an artist as the late Mr. John Brett, F.R.A.S., failed to recognise wherein they differed from an ordinary sunset, or indeed that they differed from it at all. Then when their abnormal character was recognised, the careful timing of the sinking below the horizon of the primary and secondary glows was of the first importance as giving the means for computing the height of the reflecting stratum. In the case of the appearance of such glows on future occasions, or of the crepuscular rays, or of a sun-pillar, it is very desirable that not only a precise note of the times of the phenomenon should be taken, but that also the angular extent of the various rays should not be merely roughly estimated but actually measured. A long, light, but stiff rod carrying a sliding cross-piece—a sort of tangent staff, in fact, not unlike that used by Chandrasekhara—would be of much service in determining the height and breadth of a sun-pillar, the angle which a crepuscular beam made with the horizon, the height of the brightest part of an afterglow, and so on. Such an instrument might, indeed, seem rough, but its accuracy and precision would be found to be beyond comparison greater than that of mere estimation.

CHAPTER X.

VARIABLE STARS.

IT is more than three centuries ago since David Fabricius, one of the earliest observers of sunspots, noticed that a star in Cetus, which he had observed in August, 1596, to be of the 3rd magnitude, had disappeared by October. This appeared an observation of great importance, since it seemed to show that the fixed stars are not all of them permanent, but that they might die out. Seven years later Bayer recorded a 4th magnitude star in precisely the same position as that which Fabricius had noted to have disappeared. Here, however, the matter rested for an entire generation, and it was not until 1638 that Holwarda detected the star again as of the 3rd magnitude in December, but found it disappear in the following summer to reappear again in the autumn. This star, therefore, Omicron Ceti, which received from Hevelius the name of Mira, the wonderful star, was the first to become known as a periodic variable.

The first star, that is to say, in historic times. There is another more striking even than Mira, which it seems likely was noted by the forgotten astronomers of Arabia or the valley of the Euphrates centuries before even

Hipparchus and Ptolemy compiled their catalogues. This is Beta in the constellation Perseus, described by Ptolemy as the principal star in the head of the Gorgon Medusa, which the hero is represented as carrying in his hand. This star has come down to us from the Arabs with the name Algol, the Demon Star, and it is at least a probability that it owed its name to the fact that though ordinarily of the 2nd magnitude it sinks down almost to the 4th at perfectly regular intervals of two days and twenty-one hours.

The variability of Algol was discovered in modern times by Montanari in 1669, and it was re-discovered by Goodricke in 1782. The latter observer two years later added two other variables to the list; Beta Lyræ with a period of very nearly thirteen days, and Delta Cephei, with one of five and a-third days. At this date scarcely more than a century ago these four stars were almost the only variables known to us, and variables continued to be rare objects until the middle of this century. Now their numbers have been added to so greatly that the Catalogue prepared by Prof. Chandler in 1896 comprises 400 objects, the variability of which is fairly well established, and new members of this class are being discovered every month.

The most striking star of the four with which to begin is Algol. The student, avoiding all references to Ephemerides, should look out at regular intervals and compare the brightness of Algol with certain of the neighbouring stars. Ordinarily Alpha Persei will be distinctly but not very greatly brighter than our variable, whilst Gamma, Delta, Epsilon and Zeta will be distinctly fainter. At a little greater distance are Alpha and Beta Arietis, the former slightly brighter,

the latter slightly fainter than Algol.   Alpha and Beta
Trianguli are at no great distance, and are good comparison stars when Algol has begun to fade.*

It will not be long before the observer will find that
his star is undergoing a change, and that it no longer
nearly rivals Alpha Persei or Gamma Andromedæ in
brightness.   Directly this is noticed, systematic observation should be commenced.   A star should be chosen,
reasonably near, distinctly brighter than the variable,
and a second star distinctly fainter.   It is usual among
variable star observers to estimate these differences
in " steps," these " steps " corresponding generally to
about a tenth of magnitude, though probably the beginner
will make his steps considerably larger than this.   The
central principle, however, is that two stars should
be selected, one of which the observer decides to
be clearly fainter than the variable, and the other
brighter, and yet both of them pretty near the variable
in brightness.   The student should further be careful
to record whether the difference between the variable

---

* For a list of standard magnitudes, the observer cannot do better
than procure either the "Oxford Uranometria," or the " Harvard
Photometry," either of which will give him a sufficiently complete
catalogue of the magnitudes of the "lucid " stars; that is to say, of
the stars visible to the naked eye.   There are slight differences
between the two catalogues, and, therefore, it is well that the same
catalogue should be used throughout; the two should not be used
indiscriminately.

The student will be able to practise himself in the recognition of
the value of a stellar magnitude by the study of five of the principal
stars in Ursa Minor; viz., Alpha, Gamma, Delta, Theta, and Lambda.
These rank, according to the "Harvard Photometry," as 2·2, 3·2, 4·3,
5·3, and 6·5 respectively ; in other words, they make a nearly uniform
series with a common interval of very slightly over a magnitude, and,
being all close circumpolars, are always high in the sky and ready for
comparison.

and the fainter star was equal to, greater than or less than the difference between it and the brighter. An observation, therefore, might run as follows :—

Sep. 1d. 11h. 15m. 2 > a 3 < b

where a and b are the two comparison stars. This would mean that at 11h. 15m. the variable was noted to be two " steps " brighter than a and three " steps " fainter than b; in other words, that it is slightly nearer an equality with a than with b.

If the observer practises the estimation of " steps " in the following way he will soon find it much more delicate than he could have supposed any method apart from instruments could possibly be. If two stars of equal brightness are looked at for a few seconds, first one will seem to be the brighter, then the other. But if one is very slightly the fainter of the two, it will still seem the brighter sometimes, but not quite so often as the other. Where this difference in the frequency with which one seems the brighter is just perceptible, then the difference is put as one " step." If one star is generally the brighter, but sometimes the two appear equal, the difference will be two " steps." Three " steps " will be indicated if one is always, but only very slightly, the brighter.

Of course there is no reason why the observer should confine himself to two comparison stars. To begin with, indeed, it is well that he should try more; bearing in mind that the stars should be as nearly as possible at the same altitude, as a marked difference in the height above the horizon will have a considerable effect upon the estimation.

Another caution to be borne in mind is that no effort should be made to see the variable and the comparison star at the same moment. This might mean that their images fell upon portions of the retina differently sensitive. Both stars should be looked at directly, and the eye should be turned quickly from one to the other, backward and forward, until the observer is satisfied which is the brighter, and by what amount.

Having made one set of satisfactory observations, the student should leave the star for a while—say for half-an-hour—and then make an entirely fresh set of observations. If he should be fortunate enough to hit upon the commencement of a minimum his second observation will show him the star somewhat fainter than the first, and the difference will become more marked at a third observation. The entire period of decline and recovery for Algol is nine hours, the light fading for $4\frac{1}{2}$ hours, remaining constant for a few minutes, and then gradually increasing again for another period of $4\frac{1}{2}$ hours. The light changes therefore at a most rapid rate at about $2\frac{1}{4}$ hours before minimum or about the same interval afterwards, that is to say when the change is about half completed.

The observation is a simple one, with no accessories of brilliant light or pleasing colours. Yet the young observer cannot, we think, but experience a real pleasure when for the first time his observations, carefully and systematically made and duly recorded, show him beyond a doubt that he is witnessing the dimming of the Demon Star; that he is watching across untold millions of millions of miles of space the signalling of that far distant sun. There will be a sense of achievement, greater and not less because it has been accomplished by his unaided sight, than if he had had the help of some great instrument, and if there be in him anything of the stuff of which astronomers are made, he will turn eagerly to look for other objects of study, and will wait with much interest for other opportunities of watching Algol.

He will not soon exhaust this field of work which Algol has to offer him. Minimum after minimum should be carefully watched so as to determine the period. This of course is now known with the utmost exactness, even to the thousandth part of a second, and the purpose of the student's making an independent determination is for his own training in the work, not for a closer approximation to the true elements of the star. Nevertheless it has been by the continual repetition of such observations, long after the period was precisely known, that minute variations in it have been discovered, and the student should certainly not drop Algol from his observing list until he has been able not only to work out a period for himself, and so to predict in advance future minima, but also to detect an apparent irregularity in the period which is known as "the equation of light," and which is due to the fact that light takes some 16 minutes to cross the orbit of the earth. Minima which are observed in November, therefore, when the earth is at its nearest position to Algol, come earlier than the average ; those in March and June come later.

It is of course well known now that the variability of Algol is due to its having a dark companion which revolves round it in about 69 hours. The variation in Beta Lyræ is of a more complicated kind. Here there are two minima, one less pronounced than the other, and we infer therefore that in this case both stars are bright and that they alternately eclipse each other. The variation is less than with Algol, being but little more than a single magnitude.

T

Delta Cephei has a variation of much the same amount as Beta Lyræ, but it differs from that star in that it has a slow decline and a quick recovery—the decline being 91 hours, the recovery 38.

It is, however, rather with the variables of longer period that the student will most occupy himself, and of these Mira Ceti is the most noteworthy. Its brightness at maximum varies through wide limits; sometimes it scarcely exceeds the 5th magnitude, sometimes it is distinctly brighter than the 2nd, but usually it ranks between the 3rd and the 4th. It is thus always within the range of unassisted sight at maximum, but it goes down far below that range at minimum, its faintest light bringing it down practically to the 10th magnitude. The "astronomer without a telescope," therefore can only watch it at its maxima, but these form for Mira Ceti the interesting phase. The other three stars are at all times well within the range of vision. A telescope, therefore, is not needed for them, and it is much better that it should not be used.

The four stars above mentioned are thus each the type of a different class of variation. Algol is the type of the "eclipse" stars; stars which retain their maximum light for the greater part of the time, but which, at exceedingly regular intervals, suffer a diminution of light, the diminution and the recovery taking place with the same speed. The variation in their case is not due to any change in the light-giving power of the star itself, but to the interposition of a dark satellite between the star and ourselves. Two other bright stars are known of the same class, viz., Lambda Tauri and Delta Libræ, the former with a period of nearly 4 days, the latter with one of nearly $2\frac{1}{3}$ days. From the character of their variation they are rather difficult variables to discover, but the number of known members of the class tends to increase somewhat rapidly.

Beta Lyræ, a star of "reciprocal eclipse," has no analogue amongst the brighter stars, but forms a class by itself.

Delta Cephei is a type of the "short-period variables." In these stars the light is continually varying, but the recovery is, as a rule, sharper than the decline. The period is fairly regular, extending only over a few days, and the members of the class are almost entirely confined to the neighbourhood of the Milky Way. Eta Aquilæ and Zeta Geminorum, S Monocerotis, W and X Sagittarii, S Sagittæ, and T Vulpeculæ belong to this third class.

The fourth class, that of the " long-period variables," is a very numerous one. In this class the changes of brightness are often excessive. Thus, the range for Mira Ceti is often seven magnitudes; for Chi Cygni it is eight; that is to say, the light which reaches us from the star at maximum is 1600 times as great as that we receive from it at minimum. In this class, both the variation itself, and the period in which it takes place, are apt to be irregular. Thus, as already stated, the maximum of Mira Ceti sometimes has ranked it above the second magnitude ; sometimes it scarcely exceeds the fifth. Its mean period is 331 days, but this also undergoes fluctuations. Of the stars in this class a large proportion have periods of from 10 to 14 months. Beside the two stars already mentioned, Eta Geminorum, R and U Hydræ, U Orionis, W (34) Bootis, R Lyræ, T Cygni and Rho Persei may be noted as bright members of the class. W Bootis, T Cygni, and U Orionis have periods of about a year, and R Hydræ of 425 days, whilst the period of R Lyræ is only 46 days.

The fifth class consists of the "irregular variables," stars for which no law as to the time or amount of their variation has as yet been discovered. Four notable stars

belong to this class: Alpha Cassiopeiæ or Schedar; Alpha
Orionis or Betelgeuse; Alpha Herculis or Ras al Gethi;
Beta Pegasi or Scheat.

The study of variable stars lies on the border-line of
"Astronomy without a Telescope," for though most of the
above-named stars and not a few more can be sufficiently
well observed with the naked eye, there are very many
which require considerable telescopic assistance. But the
field of work open to the possessor of a good opera-glass is
sufficiently large to satisfy the most industrious observer,
whilst as—to quote Mr. J. E. Gore, one of our highest
authorities in variable star astronomy—"instrumental aid
should be employed as sparingly as possible, a glass never
being used when the naked eye will do, or a large telescope
when a small one will serve, this research may be regarded
as essentially non-telescopic."

VARIABLE STARS FOR NAKED-EYE OBSERVATION.

Chandler's Number.	Name.	R.A. 1900	Decl. 1900	Magnitude.		Period.	Type of Variation.
				Max.	Min.		
		h. m. s	o '			d. h. m.	
209	α Cassiopeiæ	0 34 50	+55 59·3	2·2	2·8	—	Irregular.
806	o (Mira) Ceti	2 14 18	− 3 25·7	1·7-5·0	8-9·5	332	Long period.
1072	ρ Persei	2 58 46	+38 27·2	3·4	4·2	—	Irregular.
1090	β Persei (Algol)	3 1 40	+40 34·2	2·3	3·5	2 20 49	Eclipse.
1411	λ Tauri	3 55 8	+12 12·5	3·4	4 2	3 22 52	Eclipse.
1768	ε Aurigæ	4 54 47	+43 40·5	3·0	4·5	—	Irregular.
2098	α Orionis	5 49 45	+ 7 23·3	1·0	1·4	—	Irregular.
2213	η Geminorum	6 8 51	+22 32·2	3·2	3·7-4·2	231	Long period.
2375	S Monocerotis	6 35 28	+ 9 59·3	4·9	5·4	3 10 38	Short period.
2509	ζ Geminorum	6 58 11	+20 43·0	3·7	4·5	10 3 42	Short period.
3796	U Hydræ	10 32 37	− 12 51·9	4 5	6·1-6·3	—	Irregular.
4826	R Hydræ	13 24 15	− 22 45·9	3·5-5·5	9·7	425	Long period.
5274	W Bootis	14 39 2	+26 57·2	5·2	6·1	—	Long period, irreg.
5374	δ Libræ	14 55 38	− 8 7·3	5·0	6·2	2 7 51	Eclipse.
5912	g Herculis	16 25 21	+42 6·1	4·7-5·5	5·4-6·0	—	Irregular.
6181	α Herculis	17 10 5	+14 30·2	3·1	3·9	—	Irregular.
6202	u Herculis	17 13 38	+33 12·3	4·6	5·4	—	Irregular.
6363	X Sagittarii	17 41 16	− 27 47·6	4	6	7 0 17	Short period.
6472	W Sagittarii	17 58 38	− 29 35·1	4·8	5·8	7 14 16	Short period.
6733	R Scuti	18 42 9	− 5 48·7	4·7-5·7	6·0-9·0	71	Long period.
6758	β Lyræ	18 46 23	+33 14·8	3·4	4·5	12 21 47	Reciprocal eclipse.
6794	R Lyræ	18 52 17	+43 48·8	4·0	4·7	46	Long period.
7120	χ Cygni	19 46 44	+32 39·7	4·0-6·5	13·5	406	Long period.
7124	η Aquilæ	19 47 23	+ 0 44·9	3·5	4·7	7 4 14	Short period.
7149	S Sagittæ	19 51 29	+16 22 2	5·6	6·4	8 9 11	Short period.
7459	T Cygni	20 43 11	+34 0·4	5·5 ?	6 ?	365±	Long period.
7483	T Vulpeculæ	20 47 13	+27 52·5	5·5	6·5	1 11 57	Short period.
7803	μ Cephei	21 40 27	+58 19·3	4 ?	5 ?	432	Long period.
8073	δ Cephei	22 25 27	+57 54·2	3·7	4·9	5 8 47	Short period.
8273	β Pegasi	22 58 55	+27 32·4	2·2	2·7	—	Irregular.

CHAPTER XI.

The Colours of Stars.

In concluding my review of departments of observation open to the "astronomer without a telescope," I wish to glance at one which offers him some small opportunity, although in general it must be considered one for the possessors of telescopes, and those even of considerable size.

The wide difference which there is between star and star as to brightness is apparent on the very first glance towards the heavens; it requires a more careful scrutiny to realize that they differ also in their colour and in the character of their shining. The ancients carried their discrimination of the difference in the brightnesses of stars so far as to recognise six magnitudes, and added the further refinement that they noted many stars as being somewhat brighter or somewhat fainter, as the case might be, than the average star of their magnitude, but when it came to the question of colour they hardly noticed any difference at all. The stars in general were described as yellow, six only being recorded as "fiery." Of these six we should class five as being distinctly orange or red—Antares, Betelgeuse, Aldebaran, Arcturus and Pollux. The sixth, Sirius, is to us an intensely white star, and there have been many discussions as to whether it has changed its colour in the last 2000 years, or whether the description given of it— "fiery red"—is due to some mistake in the record, or whether the excessive scintillation of the star may account for it. For, as we see it now when near the horizon, a momentary flash of vivid red flames out from time to time, due to the irregular dispersion of its light in passing through the tremulous atmosphere. It is from this that Tennyson, most exact of all the poets in his scientific references, calls Sirius "fiery" in the well-known passage from the "Princess" : —

> " The fiery Sirius alters hue
> And bickers into red and emerald."

A careful comparison of star with star will soon show that this classification is far from exhausting the differences of tint which may be recognised amongst the stars visible to the naked eye. Indeed the Star Colour Section of the British Astronomical Association, under the able and energetic direction of Mr. W. S. Franks, F.R.A.S., some few years ago drew up a catalogue of all the stars of the fifth magnitude and brighter, and recognised no fewer than twenty different tints or shades. The immense majority were indeed white, or white with a more or less evident tinge of yellow, up to a fairly full yellow, and the observations were made for the most part by means of refractors of three to four inches in aperture, or of reflectors of six to eight inches. The observer, therefore, who has no telescope, or who at best possesses but a good field-glass, would neither be able to deal with stars so faint as the fifth magnitude, nor to detect as many differences of colour.

Nevertheless, with a good field-glass some three hundred stars would be within his reach, and with the naked eye alone quite one hundred, the colour of which he might successfully estimate, and in all probability with patience and experience he would succeed in grouping these into fully a dozen different categories. The work would soon be felt to be an attractive one, for delicacy of discrimination would be sure to come with practice; and the sense of the power to discriminate with quickness and certainty between two stars which at first glance showed no difference would certainly bring pleasure.

In the work of estimating the colours of stars, there are two different points to keep in mind. The one is, what point in the spectrum should be taken as best representing the dominant tint of the star. The other, what is the intensity of that tint, for it must be remembered that the stars give continuous spectra, that is to say, the groundwork of their spectra is essentially continuous. Roughly speaking they all give light of all wave-lengths; in other words, they shine essentially by white light. We have no stars with spectra limited to one particular region. Even in the banded spectra none of the seven colours that we ordinarily recognise are entirely wanting, and even if we went further, as we easily might, and divided the spectrum not into seven only, but into twelve or more different colours the same statement might hold good.

We must therefore regard the stars as shining essentially by white light. But the various tints which together go to make up a perfect white, are not always found in their exact proportion. We may regard, therefore, a coloured star as a star shining by white light plus a certain proportion of light of one or more specific colours.

Assuming that this is so, that the light of any star is partly white and partly coloured, we may divide the stars into classes, depending entirely upon the depth of tint which they show, and not upon its colour. A five-fold division suggests itself, something to the following effect :—(1) pure white, (2) tinted, (3) coloured, (4) fully coloured, (5) deeply coloured. Amongst the stars visible to the naked eye the full and deep colours are rare; it is especially in the field of double star astronomy that we get the deeper and richer tints.

After the question of the depth of tint which the stars show, comes the question of the colour of that tint. For naked-eye stars, the more refrangible colours do not come into consideration. The range is from orange-red up to yellowish green, or possibly in a single instance—that of Beta Libræ—to green. Alpha Lyræ, and possibly one or two other stars, have a distinct bluish tinge, but in general the stars not passed as white may be very well scheduled under one of the five following heads :—(1) reddish-orange, (2) orange, (3) orange-yellow, (4) yellow, (5) yellowish green.

In working upon star colours with the field-glass or naked eye it is impossible to use any artificial standard of colour, but the wide field of view, and the ease and rapidity with which the attention can be turned from one part of the heavens to the other, will much more than make up for this deficiency. The stars must be compared one with another, and the estimations of colour must be purely relative, and the method will be found much the most accurate possible.

The most satisfactory programme for an evening's work would probably be to take a number of stars not differing too widely in magnitude, and without regard to the exact name to be given to the depth of tint of each, to arrange them simply in order of colour intensity, beginning with the star of purest whiteness and going downwards to the one that showed the fullest tint. It would probably be found well for the sake of simplicity to keep this work of the arrangement in order of the depth of tint entirely separate from the work of arrangement in order of colour. For several nights a large number of stars should be taken and carefully compared, each one with the others, until a satisfactory progression has been arranged for the whole number; the depth of tint being the criterion. Then, when this work has been carried out successfully, a similar comparison should be instituted as to the colour of the stars, ranging them in order from the reddest to those which show the nearest approach to green; and in this second comparison it would probably be found useful to confine the work at first only to the stars of deep tint, then to those of full colour, and so on, until last of all those are classified according to the spectral order of their colours which are only slightly tinted. Thus, if a hundred stars have been selected as showing some more or less appreciable tint, and have been arranged in order of the depth of tint, number 100 being the star of deepest tint, then numbers 61 to 100 might be the subject of the first arrangement in spectrum order, numbers 41 to 80 of the second, and so on.

The great advantage of making the colour estimations purely comparative would be that in the case of any variation of colour in any star its detection would be easier, and much more free from ambiguity than by any other method. Thus Klein and Weber state that Alpha Ursæ Majoris passes through a series of colour changes in a period of 33 days. Several telescopic stars are supposed in like manner to have changed their colour. The case of Sirius has already been referred to, but whatever it may have been in classical times, it certainly has shown itself

unvarying white in our own day. Algol, also a white star
to us, is described by Al Sufi as "red." It is therefore
not impossible that the systematic study of star colours,
even though it be undertaken by the naked eye alone, or
at best with the assistance of a field-glass, may yet succeed
in discovering an instance of colour change. But should
it fail to do so, it must not be forgotten that the clear and
unmistakable evidence, which a thorough and systematic
record of the relative colours of stars would supply, that
no colour change had occurred, would be just as valuable,
just as important.

Closely connected with the colour of stars is their
mode of shining; in other words, their scintillation.
The blue stars and white stars, like Vega and Sirius,
"twinkle" the most rapidly; the orange stars shine the
most steadily. The planets ordinarily do not scintillate.
This scintillation is, of course, due to tremors in our own
atmosphere, since the stars are to us absolutely mathe-
matical points to which not even the most powerful
telescope has yet succeeded in giving any appreciable
diameter. It follows that when the air, through which
the star's rays pass to our eye, is in great agitation, many
rays may reach us one moment and few the next, and the
star may seem to flicker like a candle in the wind. The
planets do not scintillate because they show to us real
discs, even though these are not large enough to be
perceived as such by the naked eye. The one exception,
the planet Mercury, is an exception partly because it is
always near the horizon when seen with the naked eye,
and partly because its diameter is very small. It owes its
Greek epithet στιλβων, "glittering" or "flashing," to this
peculiarity amongst the older planets.

The measurement of the amount, or, rather, rapidity

of scintillation, lies beyond the power of the "astronomer without a telescope," but even to the naked eye there is a marked difference in this respect between star and star, and the observer will not be able to overlook its evident connection with the star's colour.

•
———

The review of the various departments of observation, available to the unassisted sight, has thus brought us to studies, as in the case of variable stars and star colours, where, though there is a certain amount of work to be done without optical aid, yet the observer's field is much increased by the possession of a good opera-glass or field-glass.

We have come as it were to the threshold of that noble structure which has been built up by means of the telescope and spectroscope ; but to that building there are many guides, and to explore it lies beyond my present task. The purpose which I have set before myself has been a much humbler one, but if I have succeeded in arousing interest in those astronomical phenomena which require no " optic glass " for their display, I shall be well rewarded.

And I think that he who seriously undertakes some department of astronomy without a telescope, will likewise not fail of his reward. The growth in the power of perception which careful practice in observation brings is a real gain. Real too is the gain in habits of system and method. For to perceive is but a part of the astronomer's work ; he must learn to record what he has perceived, and must form the habit of recording at once and recording in order. And this habit, as well as the gain in perception, means increase of power, and power gives pleasure.

Pleasure there is too in gaining as it were from direct converse with nature, fresh insight into her mysteries; pleasure, if our knowledge is really increased; pleasure, too, even if the problems prove too involved for us and our only progress be towards a truer appreciation of their difficulty.

The fields of work which we have passed in review have been both many and varied. They have extended from phenomena the most slight and transient—the lighting of a sunset cloud, the momentary flash of a meteor—to the greatest and most enduring that the universe can show— the fabric of the Galaxy and its interweaving with the stars. And there is above all in this direct study of the heavens, out in the open, beneath the deep unsounded sky, a charm and an awe, not to be realised otherwise. It is nature at her vastest that we approach; we look up to her in her most exalted form. We see unrolled before us the volume which the finger of God has written; we stand in the dwelling-place of the Most High.

INDEX.

ERRATA.

Map 4, *Constellation Bootes, for* " 69," *read* " P. xiv. 69."
Maps 5 and 8, *for* "Aquilla," *read* "Aquila."
Map 8, *for* " Sagitarius," *read* " Sagittarius."
Map 11, *for* "Pupis," *read* " Puppis."
Map 12, *for* " Columbia," *read* " Columba."

ND - #0104 - 050126 - C0 - 229/152/17 - PB - 9781330260449 - Gloss Lamination

Praise for *God is an Englishman*:

'It's a fascinating, fulsomely detailed book, which provides a wonderful tour through English Christianity.'

Rev Fergus Butler-Gallie

'At a time when more and more English people are casually abandoning Christian faith, because they no longer see its point, Bijan Omrani reminds us how much of what we are pleased to take for granted depends on it. Belief in God always requires faith, but awareness of the good things kept alive by it gives us reasons for venturing the leap. This is a beautifully composed book, and an important one.'

Rev Professor Lord Nigel Biggar

'At last someone is standing up for our Church and its amazing contribution, over centuries, to English life.'

Quentin Letts

'Druids and devotees of Woden or Thor probably lamented the rise of Christianity in England with the same sense of forsaken traditions as Christians feel today at the encroachments of secularism, godlessness, pantheism and weird fringe cults. But the loss of the country's Christian heritage would make England unrecognisable, as Bijan Omrani explains, impoverished, dreary and dim, with all that's glorious withered. With characteristically English understanding and good humour, *God is an Englishman* exposes the danger and explains why everyone in England, of all faiths and none, should relish and cherish the Christian legacy in arts and music, learning and law, kindness and kingliness, valour and values – all that's fun and all that's fundamental.'

Professor Felipe Fernandez-Armesto

Also by Bijan Omrani

Afghanistan: A Companion and Guide (with Matthew Leeming)
Asia Overland: Tales of Travel on the Trans-Siberian & Silk Road
Iran: Persia Ancient and Modern (with Helen Loveday, Bruce
Wannell and Christoph Baumer)
Caesar's Footprints: Journeys to Roman Gaul

For Children

Ticking Tom and the Tornado

God is an Englishman

Christianity and the Creation of England

Bijan Omrani

FORUM

FORUM

First published in Great Britain by Forum,
an imprint of Swift Press 2025

9 8 7 6 5 4 3 2 1

The dedication logo was drawn by Fr Owen Dobson,
and is used with his permission

Printed and bound in Great Britain by CPI Group (UK) Ltd,
Croydon CR0 4YY

A CIP catalogue record for this book is available from the British Library
We make every effort to make sure our products are safe for the purpose
for which they are intended. Our authorised representative in the EU for
product safety is Easy Access System Europe, Mustamäe tee 50,
10621 Tallinn, Estonia gpsr.requests@easproject.com

ISBN: 9781800753068
eISBN: 9781800753075

MIX
Paper | Supporting
responsible forestry
FSC
www.fsc.org   FSC® C013604

This book is dedicated to

*The Church and Congregation of St Michael and All Angels
Mount Dinham, Exeter, who distracted me frightfully whilst
I was writing this book, but to a good end.*

SEMPER FLOREAT